Reinforced Concrete from

1906 to Today

Series Editor
Gilles Pijaudier-Cabot

Reinforced Concrete from 1906 to Today

Application to the Evaluation of Structures in the Framework of an Asset Management of Civil Engineering Works

Xavier Lauzin
Liliya Shakamalova

iSTE

WILEY

First published 2025 in Great Britain and the United States by ISTE Ltd and John Wiley & Sons, Inc.

ISTE Ltd
27-37 St George's Road
London SW19 4EU
UK

www.iste.co.uk

John Wiley & Sons, Inc.
111 River Street
Hoboken, NJ 07030
USA

www.wiley.com

Library of Congress Control Number: 2025939338

British Library Cataloguing-in-Publication Data
A CIP record for this book is available from the British Library
ISBN 978-1-83669-004-7

The manufacturer's authorized representative according to the EU General Product Safety Regulation is Wiley-VCH GmbH, Boschstr. 12, 69469 Weinheim, Germany, e-mail: Product_Safety@wiley.com.

Contents

Introduction

In 1897, Charles Rabut created the first course on reinforced concrete at the École des Ponts, France. Initially composed of four lessons, the course was increased to 10 lessons in 1910. This course provided an explanation of the French ministerial regulations on reinforced concrete as well as detailed explanations of around 30 most interesting constructions.

In Chapter 2 of his 1910 course on the "General Design of Reinforced Concrete", Charles Rabut defined this construction method as "a generic form embracing an unlimited number of very different construction processes, several of which may have a broader scope than that of masonry and almost as broad as that of steel framework. The elements differentiating these systems are the relative positions of the steel and the concrete; the steel can be added to concrete (reinforced concrete), the concrete inside the metal (fretted concrete), or even incomplete penetration of one into the other, which can be achieved in various ways".

The course discussed the advantages of this new construction method, namely:

– Flexibility: it was possible to make many more different combinations with reinforced concrete than with traditional masonry or metal elements. Concrete thus offered a greater number of solutions; it could adapt better to more varied needs.

– Resistance compared to that of masonry in extension and that of metal in compression.

– Safety: Charles Rabut specified that "if unforeseen circumstances occur leading to an overturning in the direction of the forces, a masonry structure (which acts on the extension) becomes compromised and a metal structure buckles. Concrete combining the two types of resistance copes better with these unforeseen forces".

– Better resistance to dynamic loads: in particular, in terms of shocks and vibrations. This conclusion was reached following the results of a Franco–Swiss study on around 10 reinforced concrete structures in the former Jura-Simplon railway network, classified as bridges with a span of 2–5 m, built 8–15 years before. These structures with heavily loaded tracks showed no cracks.

– Better resistance to fire.

– Resistance to bad weather: the course specified that "one can wonder if the metal could be affected by humidity in cracks occurring in the concrete. If cracks occur, they always stop at the reinforcements and narrow from the external surface to the steel present, where their opening reaches zero".

– Resistance to locomotive smoke.

– Ease of execution: ease of transport of constituent materials and ease of implementation (compared to metal frames or masonry assemblies).

– Increasing stability with age of construction.

Nowadays, the engineer responsible for assessing an existing, non-damaged structure must know how to perform a structural diagnosis of the structure. First, it is necessary to ensure that the latter, subjected to loads and live loads, maintains normal behavior, that is, as considered during the design. It will therefore be necessary to determine and consider its general or local resistance according to the rules applicable at the date of construction. It is indeed necessary to know why and how the calculation and construction rules have changed, what are their inaccuracies, their inadequacies and sometimes even their errors.

The purpose of this book is to provide the different calculation codes introduced into French regulations from the first in 1906 to Eurocode 2 applicable today.

It was decided to present the different regulations as summarized by Eurocodes to facilitate understanding and comparisons:

– section 1: general overview;

– section 2: bases of design;

– section 3: materials;

– section 4: durability and cover for reinforcement;

– section 5: structural analysis;

– section 6: calculation of structures;

– section 7: constructive provisions.

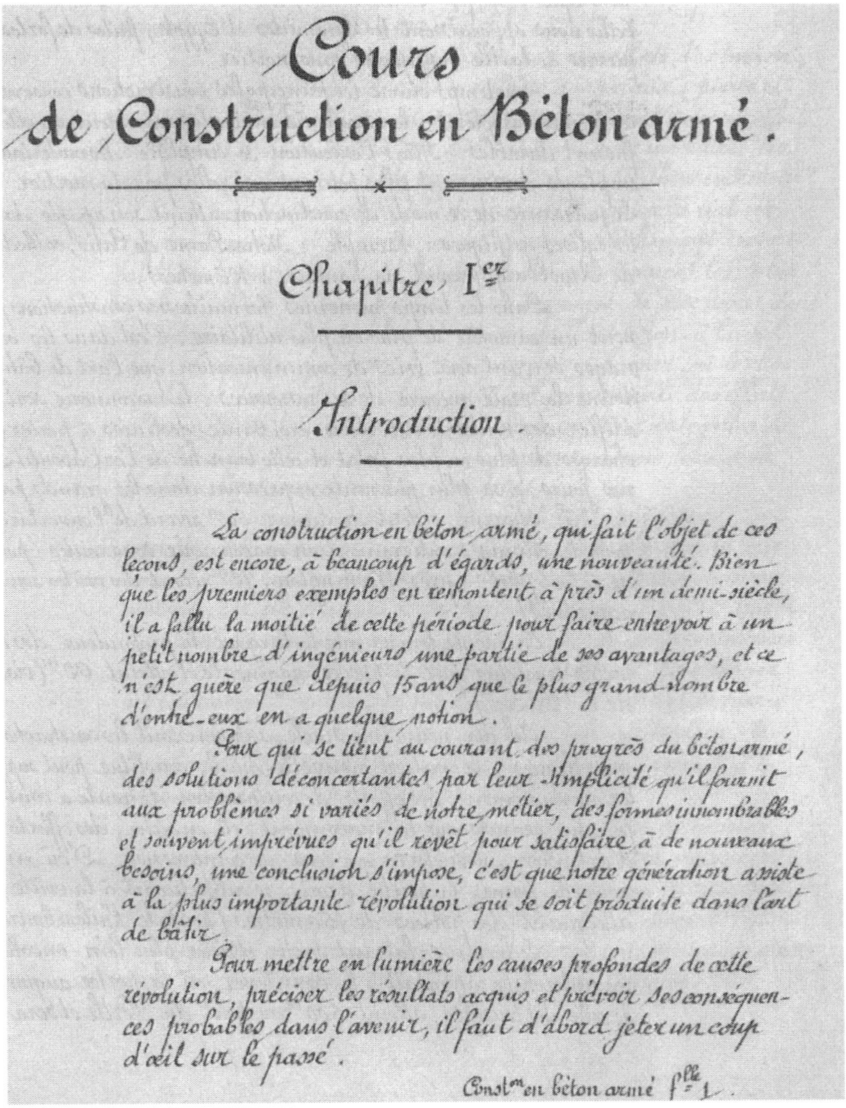

Cours
de Construction en Béton armé.

Chapitre Ier

Introduction.

La construction en béton armé, qui fait l'objet de ces leçons, est encore, à beaucoup d'égards, une nouveauté. Bien que les premiers exemples en remontent à près d'un demi-siècle, il a fallu la moitié de cette période pour faire entrevoir à un petit nombre d'ingénieurs une partie de ses avantages, et ce n'est guère que depuis 15 ans que le plus grand nombre d'entre-eux en a quelque notion.

Pour qui se tient au courant des progrès du béton armé, des solutions déconcertantes par leur simplicité qu'il fournit aux problèmes si variés de notre métier, des formes innombrables et souvent imprévues qu'il revêt pour satisfaire à de nouveaux besoins, une conclusion s'impose, c'est que notre génération assiste à la plus importante révolution qui se soit produite dans l'art de bâtir.

Pour mettre en lumière les causes profondes de cette révolution, préciser les résultats acquis et prévoir ses conséquences probables dans l'avenir, il faut d'abord jeter un coup d'œil sur le passé.

Constion en béton armé fle 1.

Figure I.1. *Reinforced concrete course (in French) by Charles Rabut*

The main dates of change in regulatory texts are summarized in Table I.1.

Year	Name	Classification	Domain						Reference	Subtitle
			Reinforced concrete	Prestressed concrete	Metal	Mixed	Civil loads	Military loads		
August 1852							X		Decree of August 10, 1852, regulating the policing of traffic and public messaging	
February 1858					X				Memorandum on railway bridge tests (for the record)	
June 1869					X		X		Memorandum of June 15, 1869, relating to the tests to be carried out on metal bridges intended for land routes	First regulation of loads and calculation for metal road bridges. Refers to the memorandum of February 26, 1858, relating to the tests of railway bridges
July 1877					X		X		Memorandum of July 8, 1877, relating to the February 26, 1858, and June 15, 1869, revisions relating to the testing of metal bridges	This memorandum deals with both the rules of loads and calculation
August 1891					X		X		Memorandum of August 29, 1891, relating to the revision of the 1877 memorandum on new rules relating to the testing of metal bridges	This memorandum, like the previous one, deals with both loads and calculations.
October 1906	BA_06	?	X				X	X	Memorandum October 20, 1906	Use of reinforced concrete
January 1915	RC_1915	15-1			X		X	X	Memorandum of January 8, 1915	Metal bridges – Load regulation
June 1920		?						X	Memorandum of June 1, 1920	
May 1927	RC_1927	27-2	X		X		X		Memorandum series A No. 3 of May 10, 1927	Metal bridges and reinforced concrete bridges
May 1930		?						X	Memorandum of May 5, 1930	
February 1933		33-1			X				Memorandum series A No. 1 of February 7, 1933	Metal bridges and reinforced concrete bridges
July 1934	BA_34	34-1	X						Memorandum series A No. 8 of July 19, 1934	Regulation BA 34
May 1937		?							Memorandum series A No. 3 of May 10, 1937	

March 1939		39-2			X				Memorandum series A No. 4 of March 7, 1939	Metal bridges and reinforced concrete bridges
May 1939		39-1						X	Memorandum of May 5, 1939	Resistance conditions of civil engineering structures
August 1940	RC_1940	40-1					X		Memorandum A-1 of August 29, 1940	Metal bridges and reinforced concrete bridges
October 1944		44-1					X		Memorandum series A No. 27 of October 14, 1944	Amendment of Article 2 of the Regulation of May 10, 1927, for the calculation and testing of metal bridges
February 1946		46-2						X	Memorandum series A No. 27 of February 11, 1946	Guidelines relating to the memorandum on heavy military equipment on civil engineering structures
February 1946		46-2						X	Memorandum series B No. 27 of February 11, 1946	Guidelines on the movement of heavy military equipment on civil engineering structures
April 1946		46-4			X				Memorandum series A No. 21 of April 30, 1946	Guidelines for the calculation of steel frames and bridges with electric arc welded joints
December 1946		46-3						X	Memorandum series B No. 243 of December 10, 1946	Guidelines on the movement of heavy military equipment on civil engineering structures
June 1947		47-1						X	Memorandum TP series B No. 110 of June 6, 1947	Conditions for the movement of heavy military equipment on civil engineering structures
March 1950		50-1						X	Memorandum TP, SP II, No. 42 of March 7, 1950	Movement of heavy equipment on civil engineering structures
October 1953	IP1_53	53-1		X					Memorandum No. 141 of October 26, 1953	Provisional guidelines for the use of prestressed concrete
April 1958	RC_1958	58-1 27-1 34-1					X	X	Memorandum No. 30 of April 5, 1958	Guidelines for live load and road bridge tests
January 1960		60-4							Memorandum SCET by M. JR Robinson (Routes review)	
August 1960	RC_1960	60-3			X		X	X	Memorandum No. 65 of August 19, 1960	Information leaflet 61, Title I to V – Design, calculation and testing of civil engineering structures

December 1962		62-2					X	X	Letter of December 8, 1962, from the Minister to the Head of the SSAR	Live load of bridges under highways
November 1964	BA_64	64-1	X						Memorandum No. 70 of November 14, 1964	Information leaflet 61, Title VI – Design and calculation of reinforced concrete structures
August 1965	IP1_65	65-1		X					Memorandum No. 44 of August 12, 1965	Provisional instruction 1 – Provisional guidelines relating to the use of prestressed concrete
March 1966	CM_66	66-1				X			Memorandum d-10944 of March 25, 1966	Medium span road bridges
December 1968	CCBA_68	68-1	X						Memorandum MEL No. 68-119 of December 11, 1968	Information leaflet 61, Title VI (CCBA 68)
February 1970	MI_70	70-2			X				Memorandum No. 70-18 of February 4, 1970	Information leaflet 61, Title V – Design and calculation of steel bridges and metal constructions
June 1970	CCBA_70	70-3	X						Decree No. 70-505 of June 5, 1970	Modification of Information leaflet 61, Title VI
October 1970	CCBA_70	70-1	X						Memorandum MEL No. 70-115 of October 27, 1970	Information leaflet 61, Title VI (CCBA 70) – Technical regulations for the design and calculation of reinforced concrete structures and constructions
December 1971	RC_1971	71-1	X	X	X	X	X	X	Memorandum No. 71-145 of December 13, 1971	Provisional instruction on common guidelines relating to the calculation of constructions
December 1971	RC_1971	71-3					X	X	Memorandum No. 71-155 of December 29, 1971	Information leaflet 61, Title II – Road bridge load and test program
December 1971	RC_1971	71-4	X	X	X		X	X	Memorandum No. 71-156 of December 30, 1971	Implementation Measures for Information leaflet 61, Title II – Highway bridge load and test program
August 1973	MI_73	73-1			X				Memorandum No. 73-150 of August 7, 1973	Information leaflet 61, Title V – Design and calculation of steel bridges and metal constructions
August 1973	IP2_73	73-2		X					Memorandum un No. 73-153 of August 13, 1973	Provisional instruction 2 – Provisional instruction no. 2 relating to the use of prestressed concrete

April 1974	IP1_74	74-1		X					Memorandum No. 74-60 of April 23, 1974	Provisional instruction 1 modified
April 1975	IP1_75	75-1		X					Memorandum No. of April 2, 1975	Provisional instruction 1 modified
February 1978	MI_78	78-1			X				Memorandum No. 78-33 of February 18, 1978	Information leaflet 61, Title V – Design and calculation of steel bridges and metal constructions
March 1979	DC_79	79-2	X	X	X	X	X	X	Memorandum No. 79-25 of March 13, 1979	Technical guidelines on common specifications dated 1979 relating to construction calculations (DC 79)
December 1979	IP1_79	doc		X					Memorandum No. 79-121 of December 14, 1979	Provisional instruction 1 modified
May 1980	BAEL_80	?	X	X					Memorandum No. 79-115 of November 2, 1979	Special Information leaflet no. 79-48 bis (BAEL 80 regulations)
May 1980	BAEL_80	80-1	X						Memorandum No. 80-70 of May 23, 1980	BAEL 80 regulations
July 1981	MI_81	81-1				X			Memorandum No. 81-63 of 28 July, 1981	Regulations for calculating steel-concrete composite bridges
October 1983	BPEL_83	83-2		X					Memorandum of October 8, 1983	Information leaflet No. 62 Title I section II (BPEL 83)
October 1983	BAEL_83	83-1	X						Memorandum of October 8, 1983	Information leaflet No. 62 Title I section I (BAEL 83)
January 1992	BPEL_91			X					Memorandum No. 92-57 of December 30, 1992	Information leaflet No. 62 Title I section II (BPEL 91)
January 1992	BAEL_91		X						Memorandum No. 92-75 of December 1, 1992	Information leaflet No. 62 Title I section I (BAEL 91)
April 1999	BPEL_99			X					Memorandum No. 99-28 of April 6, 1999	Information leaflet No. 62 Title I section II (BPEL 91 revised 99)
April 1999	BAEL_99		X						Memorandum No. 99-27 of April 6, 1999	Information leaflet No. 62 Title I section I (BAEL 91 revised 99)

Table I.1. *Classification of regulations in chronological order (source: SETRA)*

1

Material Strength

In this chapter, we only briefly discuss beam theory elements such as beams, girders, or joists and their associated assumptions:

– The material is assumed to be homogeneous and isotropic. The deformations, because of external loads, are reversible and small (linear elasticity theory).

– The movements of the material points are negligible (first-order theory).

Two laws or principles therefore result from these assumptions:

– *Generalized Hooke's law*: it states that the relationships between external forces, stresses and deformations are linear and homogeneous.

– *The superposition principle*: a stress (or strain) produced by multiple applied loads is the superposition of the stresses produced by each of the loads assumed to act in isolation.

In the case of beams, there are two additional principles:

– *Saint-Venant's principle*: the stresses in a section Σ far from the application points of the external forces depend only on the stresses of the system built by forces applied on one side only of the sigma section Σ.

– *Navier–Bernoulli's principle*: when a beam deforms, the straight sections remain flat.

Therefore, to be able to apply the beam theory, it is necessary to ensure that these hypotheses are actually respected.

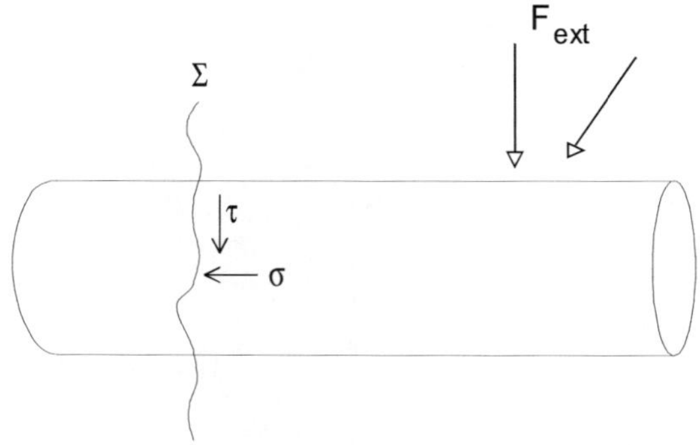

Figure 1.1. *Saint-Venant's principle*

It is generally considered that this theory gives reliable results if the following conditions are met:

– The width of the beam (cross-sectional dimension) is small compared with its length, that is, $\frac{1}{30} < \frac{h}{l} < \frac{1}{5}$ for a straight beam and $\frac{1}{100} < \frac{h}{l} < \frac{1}{5}$ for an arch, where h is the height of the beam and l is the length of the beam.

– The radius of curvature of the average fiber is greater than five times the height.

– For a beam with a variable cross-section, it must gradually vary along the mean fiber.

$\frac{1}{30} < \frac{h}{1} < \frac{1}{5}$ for a straight beam

$\frac{1}{100} < \frac{h}{1} < \frac{1}{5}$ for an arch

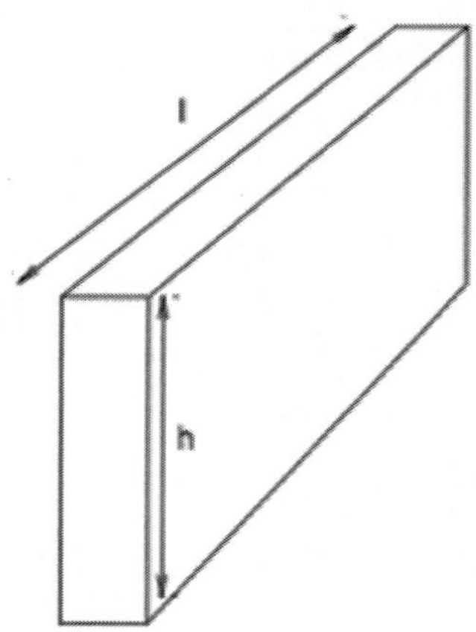

Figure 1.2. *Height/span ratio (h/l) for a beam and an arch*

The external loads applied to a beam are “actions” that produce “stresses” within the beam material. The most common stresses are as follows:

– the bending moment (bending of the beam);

– normal force (axial compression or traction);

– the shear force (beam shear).

In the following, we will study the different stresses that can be applied to beams.

NOTATIONS.–

M: bending moment

N: normal effort

T: shear force

t: shear force stress

E: modulus of elasticity

S: area of the section

σ_c: compressive stress

σ_t: tensile stress

1.1. Compression and traction

Consider a cross-section of any beam subjected to an external force perpendicular to the latter.

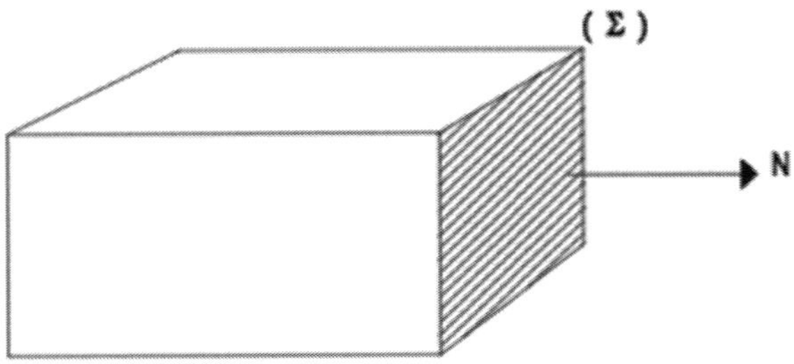

Figure 1.3. *Traction forces on a straight section*

Let S be the area of the section (Σ); this force causes a normal stress (perpendicular to (Σ)) on each element of the surface (Σ) to be constant over the entire section and equals:

$$\sigma_t = \frac{N}{S}$$

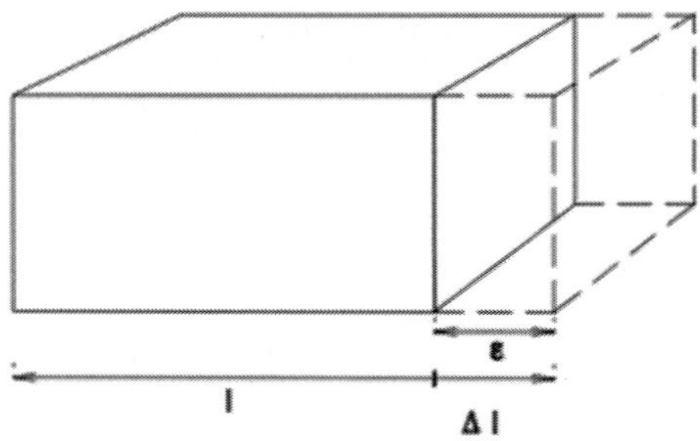

Figure 1.4. *Deformation of a prismatic element under stress by compression or traction*

Under the effect of this external force, the fibers of initial length l become elongated (case of traction) or shortened (case of compression) Δl such that:

$$\frac{\Delta l}{l} = \frac{N}{ES}$$

Indeed, according to the generalized Hooke's law, we know that the deformation equals:

$$\varepsilon_t = \frac{\sigma_t}{E} \text{ or } \varepsilon_t = \frac{\Delta l}{l} \text{ and } \sigma_t = \frac{N}{S}$$

Hence:

$$\frac{\Delta l}{l} = \frac{N}{ES}$$

The deformation therefore causes elongation for a traction effect and shortening for a compression force.

1.2. Pure flexion

Consider any beam and two straight sections of this beam.

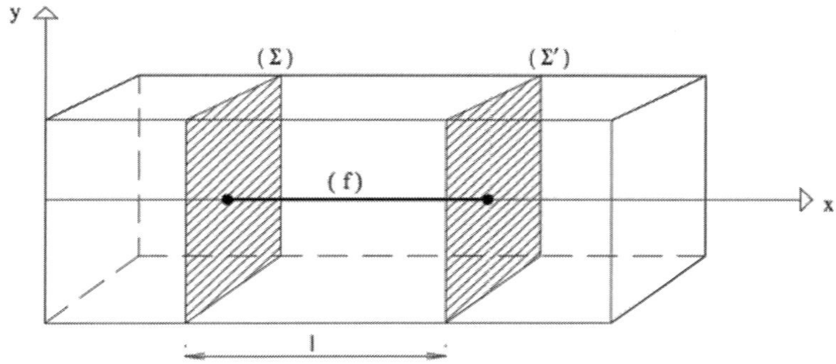

Figure 1.5. *Cross-sections of a beam being bent*

If we subject this beam to a system of forces generating a bending moment, according to the Navier–Bernoulli principle, the sections (Σ) and (Σ′) remain straight after deformation.

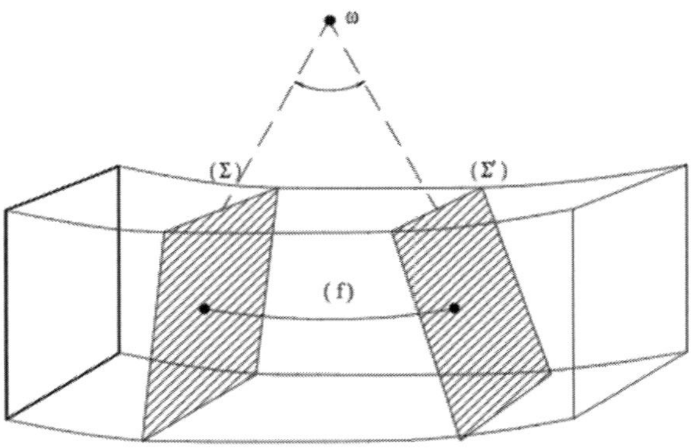

Figure 1.6. *Cross-sections after application of the bending moment*

The elongation of any fiber (f) between sections (Σ) and (Σ′) is a linear function of its coordinates in section (Σ).

The stress in the fiber (f) is then determined by considering Hooke's law:

$$\sigma = a + by + cz$$

where the constants a, b and c are determined by the equivalence principle:

$$\iint_{(\Sigma)} \sigma dy\ dz = 0 \qquad \text{(force equilibrium)}$$

$$\iint_{(\Sigma)} \sigma . y . dy\ .dz = M \qquad \text{(moment equilibrium)}$$

$$\iint \sigma . z dy\ dz = 0 \qquad \text{(we assume M is directed along x)}$$

This ultimately leads to:

$$\sigma = \frac{M \quad y}{I}$$

where I is the inertia of the section relative to the main axis with the bending moment.

The elongation of the fiber (f) is then written as:

$$\varepsilon = \frac{\Delta l}{l} = \frac{M\ .\ y}{EI}$$

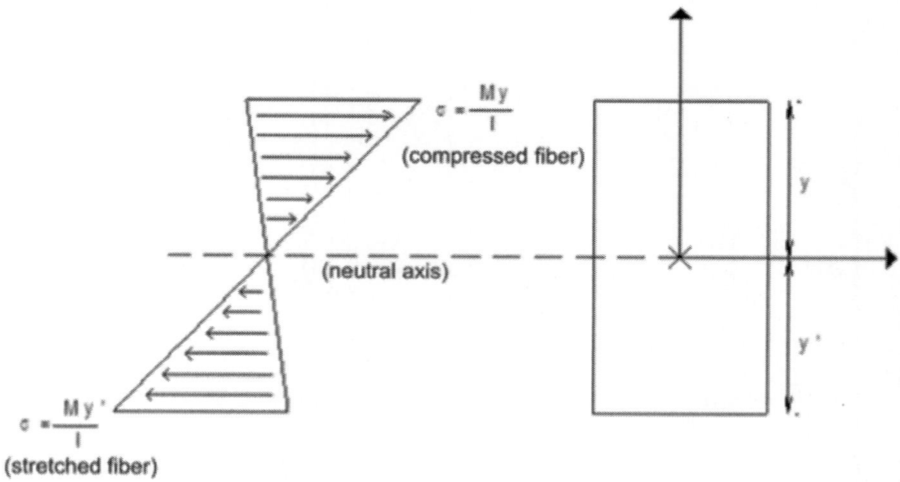

Figure 1.7. *Stress diagram of a cross-section of a beam subjected to a bending moment*

The relative movement of (Σ) and (Σ′) therefore results in a rotation:

$$\omega = \frac{M.L}{EI}$$

The constraints are represented in the diagram in Figure 1.7.

1.3. Shear strain

The effects of the shear force and bending moment usually occur simultaneously. The shear strain is given by:

$$t = \frac{T_m}{l_b}$$

where T_m is the static moment of the area above parallel to Gz. The maximum shear strain is given by:

$$t\ max = \frac{T}{bZ}$$

$$z = \frac{1}{\mu_0} \text{ internal torque lever arm}$$

The shear strain is then:

$$\gamma = \frac{T}{GS_I}$$

where S_I is the reduced section.

For:

– a rectangle: $S_I = 5/6\ S$;

– a diamond: $S_I = 30/31\ S$;

– a circle: $S_I = 9/10\ S$.

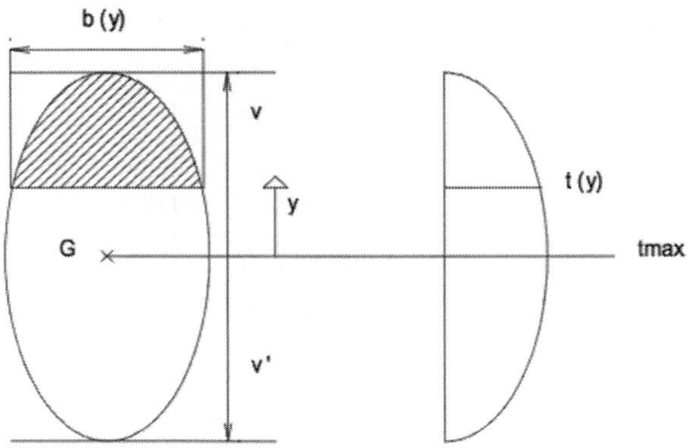

Figure 1.8. *Diagram of shear stress on a straight beam subjected to a bending moment*

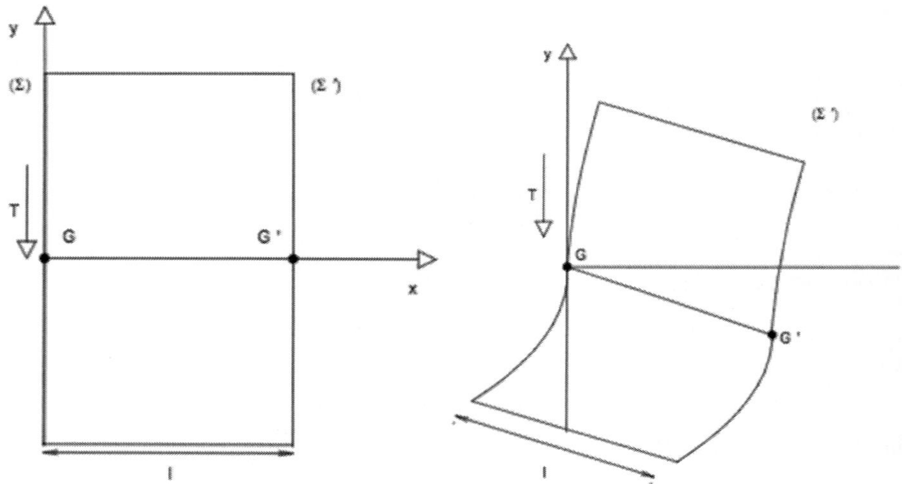

Figure 1.9. *Strain diagram*

1.4. Torsion

Studying torsion is relatively complex and is often done using the theory of elasticity, from which we retain the following elements:

– Torsional stresses are tangent stresses that are superimposed on the shear.

– These constraints are perpendicular to the radius vector coming from the center of torsion.

– For a circle, these constraints are proportional to this radius.

1.4.1. *Circular section*

When the deformation of a section due to torsion is analyzed, point M′ becomes M'_1 and therefore:

$$M'M'_1 = \gamma\ dx = \tau/G\ dx = \rho\ d\Theta$$

where $\rho = GM = G'M'$.

That is, $\tau = G\rho\ d\Theta/dx$.

The elemental force $\tau\ d\varpi$ then produces an elementary moment $\rho\tau\ d\varpi$. The sum of these elementary moments must balance the torsional moment Mt, hence:

$$Mt = GI_p\ d\Theta/dx = \tau/\rho\ I_p$$

where I_p is the polar moment of inertia.

We deduce that:

$$\tau = \rho\ Mt/I_p \text{ and } d\Theta/dx = Mt/(GI_p)$$

1.4.2. *Applying this to a circle*

Applying the polar inertia calculation to a circle gives the following result:

$$Ip = Ix + Iy = 2\ \pi\ R^4/4 = \pi\ R^4/2$$

That is, $\tau_{max} = 0.637\ Mt/R^3$, and $d\Theta/dx = 0.637\ Mt/(GR^4)$.

1.4.3. *Rectangular section*

It is calculated from a series development and a formula of the same type as for the circular section:

$$d\Theta/dx = Mt/(G*J)$$

where J is the torsion modulus, $J = Kab^3$.

We deduce that:

$d\Theta/dx = Mt/(GKab^3)$ and $\tau_{max} = K'\,Mt/(ab^2)$.

Caquot's approximation is used to determine parameters K and K′.

If $m = a/b$:

$$1/K = (1 + 1/m^2)[0.225 - 0.035\,((m - 1)/(m + 1))^2]$$

$$K' = 0.601 - 0.226\,(m - 1)/(m^2 + 1)^{0.5}$$

1.5. Spherical shell theory

Considering a spherical shell of radius r, opening radius b in horizontal projection and deflection f, the radius of the sphere is then:

$$r = (b^2 + f^2)/2f$$

The angle at the vertex is equal to $\varphi = \pi/2 - \alpha$ according to the diagram in Figure 1.10.

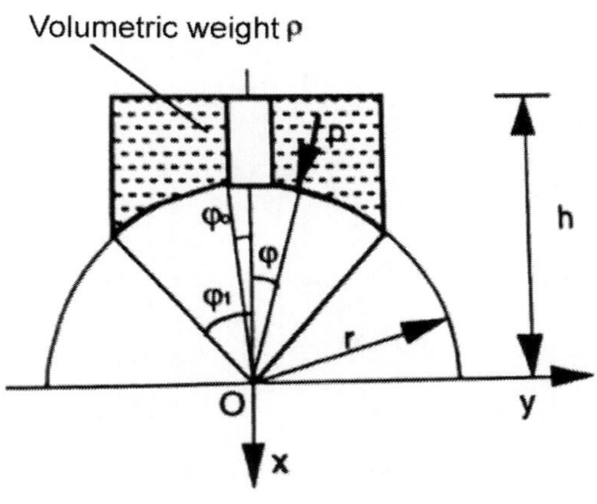

Figure 1.10. *Spherical shell diagram*

With these notations, $x = -r\cos\varphi$ and $dx = r\sin\varphi\,d\varphi$; $y = r\sin\varphi$ and $dy = r\cos\varphi\,d\varphi$; and $ds = r\,d\varphi$.

The equation is $d(\sigma\ r\ \sin^2\varphi)/r\ d\varphi = r\ \sin\varphi\ p_x/e$, where p_x is the component according to x of the load p applied to the shell and e is the thickness of the shell.

The normal stress on the lower edge parallel to the meridian is σ.

The second equation is then $d(\sigma\ r\ \sin\varphi\ \cos\varphi)/r\ d\varphi = r\ \sin\varphi\ p_y/e + \sigma_\theta$, where p_y is the component according to y of the external load.

The normal stress on the lower edge along a parallel equals σ_θ.

Both equations give $\sigma_\theta = r/e\ (p_x \cos\varphi - p_y \sin\varphi) - \sigma$.

– If we apply a self-weight load g, the formula becomes the following, using the notations in Figure 1.10:

$$\sigma = \sigma_0 \sin^2\varphi_0/\sin^2\varphi + gr/e\ (\cos\varphi_0 - \cos\varphi)/\sin^2\varphi$$

$$\sigma_\theta = gr/e \cos\varphi - \sigma$$

In the case where there is no opening at the top of the dome, then $\varphi_0 = 0$, and the equations become:

$$\sigma = gr/e\ (1 + \cos\varphi) \text{ and } \sigma_\theta = gr/e\ (\cos\varphi - 1/(1 + \cos\varphi))$$

– If we apply a load p per horizontal m^2 (snow, for example), the formula becomes the following, using the notations in Figure 1.10 and $p_x = p \cos\varphi$:

$$\sigma = \sigma_0 \sin^2\varphi_0/\sin^2\varphi + pr/e\ (1 - \sin^2\varphi_0/\sin^2\varphi)$$

$$\sigma_\theta = pr/e \cos^2\varphi - \sigma$$

In the case where there is no opening at the top of the dome, then $\varphi_0 = 0$, and the equations become:

$$\sigma = pr/e \text{ and } \sigma_\theta = pr/e\ (\cos^2\varphi - 1/2)$$

– If we apply a charge q per vertical m^2 (live load, for example), the formula becomes the following, using the notations in Figure 1.10 and $p_y = -q \sin\varphi$:

$$\sigma = \sigma_0 \sin^2\varphi_0/\sin^2\varphi$$

$$\sigma_\theta = qr/e \sin^2\varphi - \sigma$$

In the case where there is no opening at the top of the dome, then $\varphi_0 = 0$, and the equations become:

$$\sigma = 0 \text{ and } \sigma_\theta = qr/e \sin^2\varphi$$

– If we apply a uniform pressure p (gas pressure, for example), the formula becomes the following, using the notations in Figure 1.10 and $p_x = p \cos\varphi$ and $p_y = -p \sin\varphi$:

$$\sigma = \sigma_0 \sin^2\varphi_0/\sin^2\varphi + pr\,(1 - \sin^2\varphi_0/\sin^2\varphi)/2e$$

$$\sigma_\theta = pr/e - \sigma$$

In the case where there is no opening at the top of the dome, then $\varphi_0 = 0$, and the equations become:

$$\sigma = pr/2e \text{ and } \sigma_\theta = pr/2e$$

– If a hydraulic load with density ρ is applied (water load on the bottom dome of a reservoir, for example), the formula becomes the following, using the notations in Figure 1.10 and $p_x = p \cos\varphi$ and $p_y = -p \sin\varphi$ and $p = \rho(h + x) = \rho(h - r \cos\varphi)$:

$$\sigma = \sigma_0 \sin^2\varphi_0/\sin^2\varphi + \rho rh/2e\,(1 - \sin^2\varphi_0/\sin^2\varphi) + \rho r^2/3e\,(\cos^3\varphi - \cos^3\varphi_0)/\sin^2\varphi$$

$$\sigma_\theta = \rho r/e\,(h - r \cos\varphi) - \sigma$$

In the case where there is no opening at the top of the dome, then $\varphi_0 = 0$, and the equations become:

$$\sigma = \rho rh/2e + \rho r^2/3e\,(\cos^3\varphi - 1)/\sin^2\varphi \text{ and } \sigma_\theta = \rho r/e\,(h - r \cos\varphi) - \sigma$$

1.6. Cylindrical shell theory

Taking a cylindrical shell of radius r and thickness e, the shell equations are simplified by considering that:

$$y = r;\ dx/ds = 1 \text{ and } dy/ds = 0$$

The equations become $\mathbf{d\sigma/dx = p_x/e}$ and $\mathbf{\sigma_\theta = -r\,p_y/e}$.

– If we apply a self-weight load g, the formula becomes:

$\sigma = \sigma 0 + gx/e$, where x denotes the shell height above the calculated point

$\sigma_\theta = 0$

– If a hydraulic load with density ρ is applied (water load in a reservoir, for example), the formulas become, with $p_x = 0$ and $p_y = q$:

$\sigma = \sigma_0$

$\sigma_\theta = qr/e$

2

Concrete

2.1. Overview

It is common to attribute the discovery of cement to the Romans. They were probably the first to mix volcanic ash from the Pozzuoli region with lime and discovered that the mixture hardened in the presence of water. They were thus able to use it in masonry to bind stones together. At that time, the “mason trowel” appeared.

Curiously, this discovery remained dormant in the Middle Ages and only reappeared in the 18th century, in 1756, to be precise, in work by the English engineer John Smeaton. He rediscovered the properties of clay in limestone. The lime burners Parker and Wyatt industrialized this cement in approximately 1786.

In France, it was not until 1817 that the work of Louis Vicat highlighted a theory on the hydraulicity of lime and mortars.

In 1824, the Scottish engineer Joseph Aspdin filed a patent for “Portland cement”, which was similar in color to the stone found in quarries in the Portland Peninsula.

In 1833, the polytechnician Pavin de Lafarge built lime kilns at Teil, France and the first cement factory was created by Dupont and Demarle in Boulogne-sur-Mer, France, in 1848.

Finally, in 1855, the French architect François Coignet built the first concrete building on Rue Danton in Paris.

At the same time, steel production remained relatively low until the English began using coal and then coke in blast furnaces. Thus, real steel production began with the invention of the Bessemer furnaces in 1855.

Renowned for its compressive strength qualities, cement revealed its weaknesses in other areas of use. This gave rise to the idea of combining cement and metal in 1845, which became known as reinforced cement, the first example of which was Lambot's boat presented at the 1855 World's Fair in Paris.

The development of reinforced cement, followed by reinforced concrete, increased and saw significant examples in work by the French engineer François Hennebique from 1879 onward. The latter gave rise to the invention of structures resembling wooden frames but entirely made of reinforced concrete.

As early as 1896, projects for prefabricated houses made of reinforced concrete appeared.

At the beginning of the 20th century, traditional architecture was turned upside down by the discovery of this new material. On October 20, 1906, the first "ministerial guidelines relating to the use of reinforced concrete" appeared. These regulations continue to evolve according to the new characteristics of the two main components of reinforced concrete: steel and cement. This is why, when assessing an existing structure, it is essential to know the approximate age of the construction as well as the regulations applicable at that time. In particular, the load-bearing capacity of reinforced concrete elements must be assessed according to the regulations in force on the date of construction.

NOTE.– Among the precursor structures from 1855 onward, the first floor was constructed with profiled reinforcement elements coated with concrete by François Coignet. A little later, Joseph Monier replaced the wooden boxes of orange trees in the palace of Versailles with boxes made of reinforced concrete ("mobile pond boxes applicable to horticulture", patent filed for 15 years). He also built a 130 m^3 reservoir in Bougival, France. The patents filed by Monier were rarely applied for in France but were later sold to Germany, where they found popularity through the company Freytag (Wayss und Freytag). It was mainly under the impetus of François Hennebique that reinforced concrete took off in France with the appearance of "construction systems" (Hennebique, Coignet, Cottancin, Boussiron, Piketty, etc.). These systems differ from each other in terms of the calculation methods used and the construction arrangements (reinforcements, etc.). In 1874, François Coignet saw his concrete used in the construction of the Cleft Ridge Bridge in New York. These empirical methods of calculation were replaced at the beginning of the 20th century by scientific methods under the leadership of Edmond Coignet and Napoléon de Tedesco (1894).

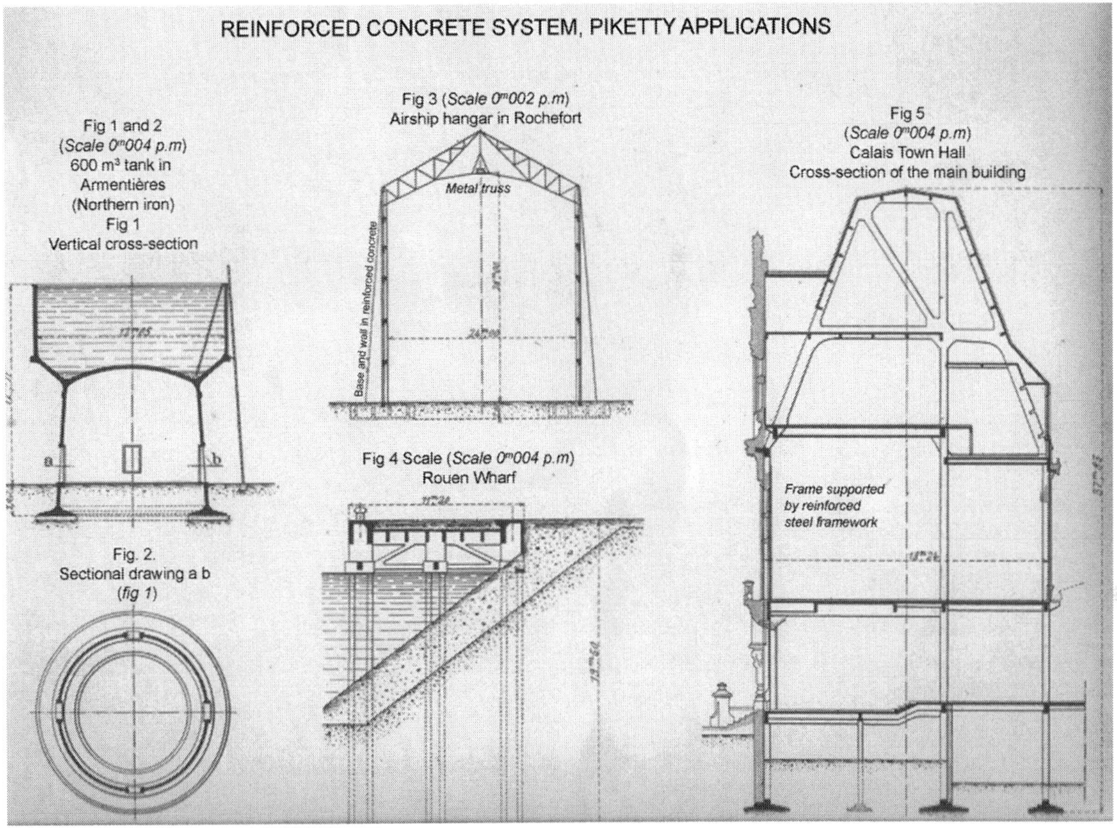

Figure 2.1. *Pikettys method*

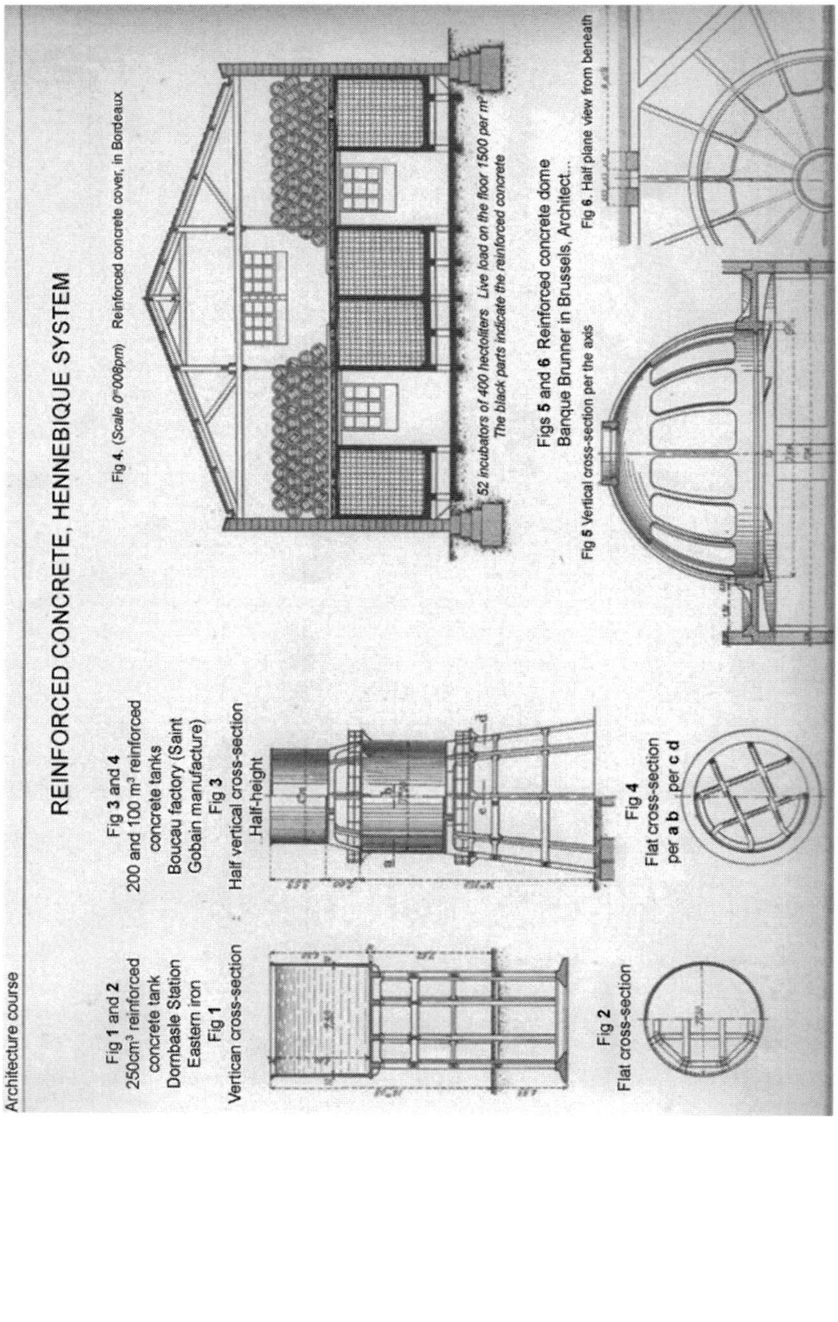

Figure 2.2. (A) *Hennebique method*

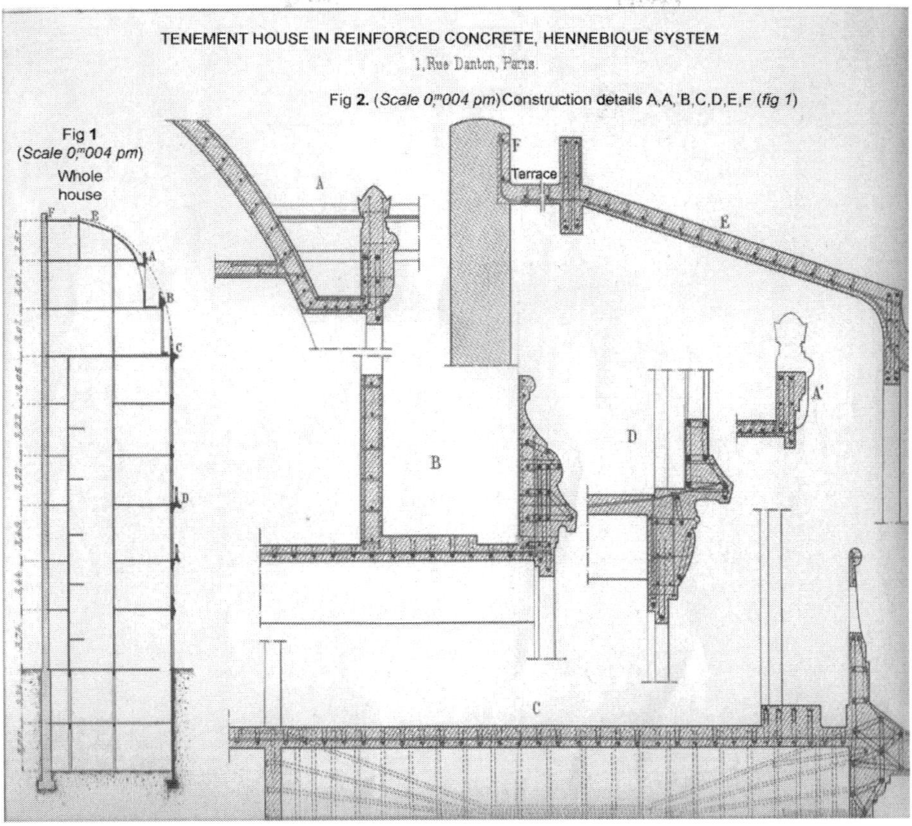

Figure 2.2. (B) *Hennebique method (continued)*

In 1906, most European countries were less advanced than France was on the subject. In the United States, reinforced concrete, which required a lot of labor, was considered less advantageous than other construction systems (metal in particular). However, its use with fewer but larger diameter reinforcements (meaning limited labor time) was of interest. The American builders used what was called Thacher-type "printed reinforcements", which provided additional anchoring to compensate for the lack of adhesion. The concretes used were more fluid because they were not loaded, which was typical in Europe. Most structures built with reinforced concrete in the United States were bridges per the Strauss system (Belvidere Bridge, for example). The voussoirs were made of reinforced concrete, prefabricated and assembled on site.

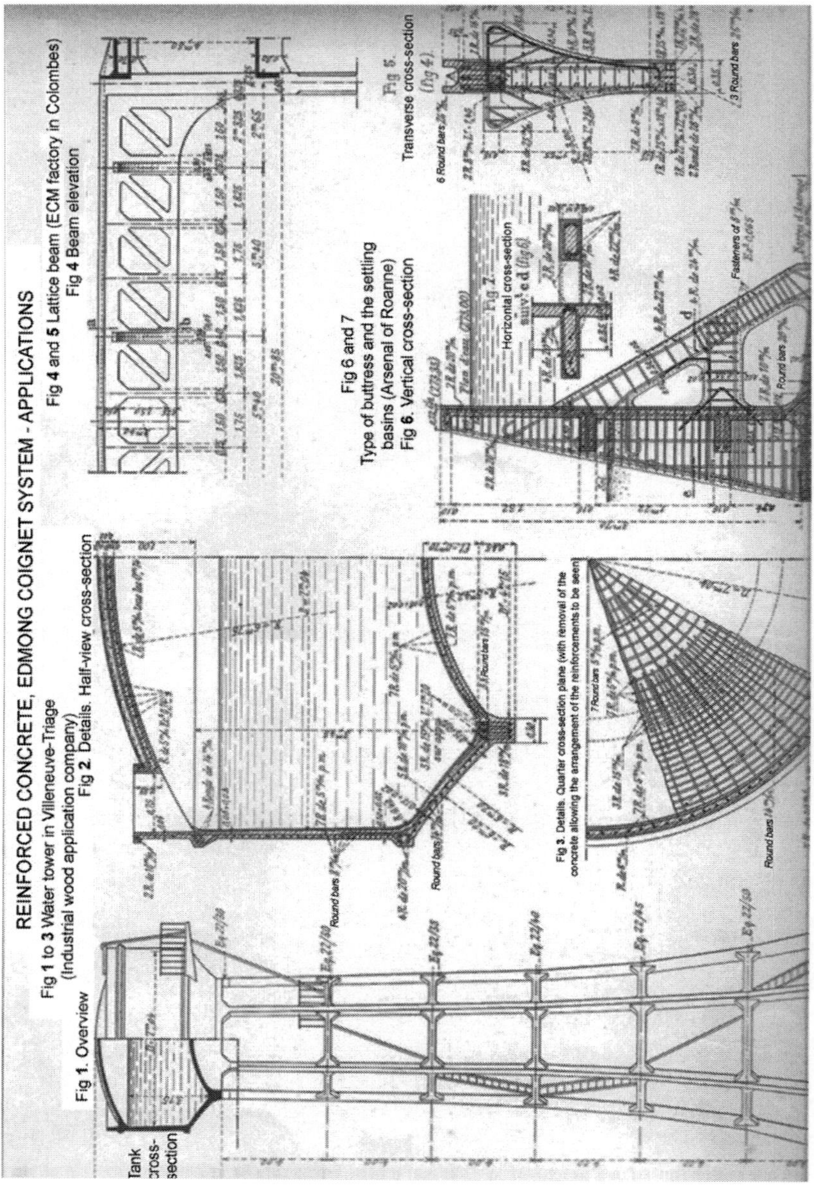

Figure 2.3. *Edmond Coignet method*

Notably, the chronological order in which the different types of applications of reinforced concrete appeared reveals the reasons that accelerated or delayed its development. The first samples were taken during the construction of buildings. Individuals had complete freedom with their own constructions, which explains the rapid progress of reinforced concrete in industrial buildings. In most countries, regulations were then created that allowed the development of railways, sanitation (an important example being the city of Paris) and, later, ports and bridges.

2.2. The composition of the concrete

Concrete is a homogeneous mixture of the following components.

2.2.1. *Cement*

Cement is composed of a mixture of limestone and clay and is ground at high temperatures into a fine powder. With water, this mixture forms a paste that can "set" and gradually harden (hydraulic binder).

The type of cement and its dosage depend on the desired mechanical performance, the resistance to possible aggressive agents and the nature of the other components present. There are generally five main categories of cement:

– clinker-based Portland cements (CPA);

– clinker-based Portland compound cements (CPJ) with the addition of other components;

– slag-based blast furnace cements (CHF and CLK) (slag is a mineral residue from the preparation of cast iron in blast furnaces);

– pozzolanic cements (CPZ);

– slag and ash cements (CLC) based on clinker, fly ash and slag.

In addition to these five major categories, there are other families of cement whose use is not permitted in reinforced concrete (hydraulic lime, XHN, etc.).

2.2.2. *Aggregates*

Aggregates are inert materials (sand, gravel, pebbles, etc.) in the composition of concrete. A distinction is generally made between natural aggregates (rolled or crushed) and artificial aggregates (industrial by-products such as crushed crystallized slag or granulated slag).

Aggregates are traditionally considered to be the base of concrete. Knowing their physicochemical and mechanical characteristics, as well as their suitability for concrete, is important.

Notably, a variation in the cleanliness of the aggregates (e.g. lack of washing) can explain a decrease in the mechanical strength of the concrete by approximately 10–15% for identical compositions. This is not directly linked to the presence of the fine materials themselves but to the fact that these fine elements impair the bonding of the cement paste to the gravel surface.

2.2.3. *Mixing water*

It allows the hydration of a mixture and is defined by its physical and chemical characteristics. It gives the mixture its fluidity (rheology).

2.2.4. *Adjuvants*

Also known as admixtures, these are products used in small quantities that can improve certain properties of concrete. They can, for example, act on:

– the setting time;

– mechanical characteristics;

– waterproofing;

– the implementation.

In the context of existing structures, from 1909 onward, sugar was used as a setting retarder. Between 1910 and 1920, products on the market based on calcium chloride (water repellent and setting accelerator) appeared.

In 1964, the COPLA (Standing Committee on Hydraulic Binders and Concrete Admixtures) was created, which was responsible for establishing a list of products that could be used safely. Later, in 1984, certification was set up using the NF adjuvants mark. The current EN 934 standard classifies admixtures into three main categories:

– those that modify the workability of concrete;

– those that act on setting and hardening;

– those that modify certain mechanical properties.

More recently, fibers have been used to reinforce the action of traditional reinforcements by opposing the propagation of microcracks. Today, we find glass fibers, metal fibers and polypropylene fibers on the market.

Currently, reactive powder concretes (RPCs), the latest products from Bouygues' scientific department, are being tested.

By analogy with the family of high-performance concretes (HPCs), research aims to improve the homogeneity of the material and its compactness.

2.2.5. *The mechanical characteristics of concrete*

In this section, we focus on the final characteristics of the concrete.

2.2.5.1. *Strength*

This is the most important characteristic of concrete. The strength is an increasing function of the cement/water ratio and compactness. By definition, concrete has good compressive strength but very low tensile strength. The compressive strength value is generally given at 28 days and is measured during destructive tests on cylindrical samples at break. Because of this low tensile strength (a crack can negate all strength), it was decided from the first regulations (1906) onward to consider only the compressive strength of the concrete in the calculations.

For example, in the 1930s, a normally dosed gravel concrete with 350 kg of Portland cement could have a compressive strength of 182 kg/cm^2 at 90 days (18 MPa). In the 1960s, normally dosed concrete (350 kg/m^3 of CPA) reached compressive strength values of 725 kg/cm^2 at 28 days (72.5 MPa), whereas reactive powder concrete was found to have values of 2,000–8,000 kg/cm^2 (200–800 MPa).

2.2.5.2. *Shrinkage*

It is a size reduction phenomenon that accompanies the setting concrete. Georges Dreux, in his *Nouveau Guide du béton*, attributed shrinkage to a decrease in temperature.

Experimental studies have shown that hardening concrete in water greatly reduces the effects of shrinkage. This is why it is common to water (or apply curing agents) to concrete parts during their curing phase when the concrete has only a low tensile strength and can crack easily under the effect of shrinkage.

Subsequent regulations specified the conditions under which there was no need to consider the effects of shrinkage (and temperature variations) or specify their values for calculations. For example, in 1932, experiments by Mr. Pugnet published in *Annales des ponts et chaussées* highlighted tensile forces in concrete of approximately 3–15 kg/cm² (0.3–1.5 MPa) depending on the percentage and storage conditions. It was then necessary to consider these factors, along with those caused by permanent loads, live loads, etc., when designing concrete elements. Later, the BAEL 83 regulations set construction lengths for which the effects of shrinkage and thermal expansion could be omitted in calculations.

For example, the shortening of concrete due to shrinkage in France is around:

$$\mathbf{\Delta l/l} = 3 \times 10^{-4}$$

This shortening results in tensile stresses such as:

$$\Delta l/l = \sigma_b/E_b$$

where:

– σ_b represents the tensile stress of the concrete due to shrinkage alone;

– E_b represents the concrete strain modulus, where:

$$\sigma_b = \Delta l/l.E_b = 60 \text{ bar} = 6 \text{ MPa}$$

However, when the tensile strength of the concrete is much lower than 60 bar (approximately 20 bar), the concrete is cracked, which justifies the assumption that the tension of the concrete in the calculations is not considered.

2.2.5.3. *Creep*

This is a phenomenon of delayed strain in concrete under constant load.

By comparison, we can use old wooden floors where the main beams are often under significant strain. For concrete, beyond a certain load (approximately half of the ultimate compressive strength), the concrete behaves plastically. Even after the load is removed, strain remains. This strain, owing to the creep of the concrete, continues for several months or even years.

Creep (like shrinkage) is related to the cement composition. Generally, it is possible to reduce creep by reducing the quantity of the cement mixture, which requires increasing the quantity of aggregates. However, a minimum cement composition must be maintained to meet other constraints (strength, durability, etc.).

2.2.5.4. *Thermal expansion*

The thermal expansion of concrete is characterized by a coefficient that depends mainly on that of the aggregates (sand and gravel), which occupy 70% of the volume of the concrete. This coefficient can vary, depending on the nature of the aggregates, from 6×10^{-6} for limestone aggregates and 13×10^{-6} for siliceous aggregates.

A thermal expansion coefficient of concrete of 1×10^{-5} is generally accepted. For France, a temperature variation of $\Delta\Theta = \pm 20°C$ is commonly accepted, which implies a variation in length:

$$\Delta l/l = 2 \times 10^{-4}$$

2.2.5.5. *The strain modulus E*

Derived from the theory of elasticity for which the strain is proportional to the applied stresses, the strain modulus (or coefficient of elasticity) is defined by:

E = unit stress/relative strain

Considering the above (strain due to creep), we consider two strain moduli, *an instant modulus* and *a deferred modulus*, to consider the fact that the total strain (including the effects of creep) is approximately three times greater than the instantaneous strain.

This longitudinal strain is accompanied by a transverse strain called the "Poisson effect". Poisson's ratio (the ratio between the transverse strain and the longitudinal strain) has a value that is generally considered equal to 0.2.

In practice, the elasticity modulus of concrete depends largely on the value of the elasticity modulus of the aggregates, which make up approximately 70% of the mixture. The choice of aggregates therefore has a significant influence on the value of the modulus for the same compressive strength.

2.2.5.6. *Traction–stress diagram*

It translates the mode of traction in the concrete as a function of the stress applied to the sample.

We previously reported that the traction modulus (or elasticity coefficient or Young's modulus) measured the capacity of concrete to deform under stress. For example, in the 1935 regulation, concrete with a compressive strength of 250 kg/m^2 was considered to break under traction of 20 kg/cm^2.

2.2.5.7. *Tensile strength*

Owing to its intrinsic composition, concrete has a low tensile strength, approximately 10–12 times less than its compressive strength.

3

Ministerial Guidelines of October 20, 1906, on the Use of Reinforced Concrete

3.1. Overview

Following the creation of the first course on reinforced concrete in 1897 at the École des ponts et chaussées, Paris, the French Ministry of Public Works published initial guidelines for calculating reinforced concrete constructions. The French Commission on reinforced cement and concrete, created after the 1900 Paris Exposition, was tasked, under the presidency of the General Inspectorate of Bridges and Roads (M. Lorieux), with establishing scientific guidelines allowing the use of this new material. This commission included experts such as Armand Considère, Charles Rabut, Augustin Mesnager, Edmond Coignet and François Hennebique.

For 5 years, they carried out an extensive testing program to better understand the mechanical properties of steel and concrete. Since unanimity was not achieved upon interpreting the test results, the commission was replaced by a new commission chaired by Mr. Lévy. The 1906 resulting guidelines were applied to buildings as well as civil engineering structures (bridges, etc.), but they were only mandatory for structures under the authority of the French Ministry of Public Works.

It comprises 25 articles distributed as follows:

– articles 1–3: live loads;

– articles 4–8: work limits;

– articles 9–13: strength calculations;

– articles 14–20: erection work;

– articles 21–25: testing.

It was completed in 1912 by a note on the calculation of a reinforced concrete hollow floor. The following symbols were adopted:

– Rb: compressive stress of the concrete;

– Ra: compressive stress of steels;

– N: compressive force;

– Ω: cross-sectional area;

– Ωa: steel cross-sectional area;

– Ωb: concrete cross-sectional area;

– Ea: modulus of elasticity of the metal;

– Eb: modulus of elasticity of the concrete;

– m = Ea/Eb: density;

– M: bending moment;

– I t: moment of inertia of the total section.

3.2. Calculation bases: requirements

3.2.1. *Basic requirements*

The following principles were considered:

– the calculation of "allowable stresses" was carried out;

– characteristic values were used.

Loads were combined without a partial coefficient. It was requested that the greatest live loads that the structure will have to support be considered (for buildings, reservoirs, pipes, etc.).

Bridges were subjected to the live loads of metal bridges according to the regulation of August 29, 1891 (replaced by that of May 10, 1927).

3.2.2. *Project usage duration*

The project usage periods were not mentioned in the directive.

3.2.3. *Thermal effects*

There are no guidelines on the possible thermal effects and how to take them into account. *Article 9* specifies, however, that "thermal efforts and those of concrete shrinkage [will have to be taken into account] whenever the free expansion of works is not provided".

3.2.4. *Settlements and differential movements*

There are no specific data on this subject in the directive.

3.2.5. *Material properties*

The directive does not mention any specifications for taking into account the creep and shrinkage of concrete.

3.3. Materials

3.3.1. *Concrete*

3.3.1.1. *Resistance*

Article 4 of the directive limited the value of the mechanical resistance of concrete to compression ("fatigue" limit) to:

28/100 × crush resistance

Notably, the concrete test specimens were 20 × 20 × 20 cubes, and they were crushed at 90 days.

NOTE.– This value was significantly higher than that of foreign regulations, for which it was common to take ¼ of this crushing value at 28 days. The French Commission of Reinforced Cement tested concrete with the following compositions:

– 400 L of sand;

– 800 L of gravel;

– 250–600 kg of Portland cement.

The results are summarized in Table 3.1.

Amount of cement (kg/m³)	Crush resistance at 28 days (MPa)	Crush resistance at 90 days (MPa)	Fatigue limit resistance (1/4 of the break at 28 days) in MPa	Fatigue limit resistance (28/100 of the break at 90 days) in MPa	Variation	Ultimate shear strength of concrete in MPa
300	10.7	16	2.7	4.4	63%	0.45
350	12	18	3	5	67%	0.50
400	13.3	20	3.3	5.6	70%	0.56

Table 3.1. *Concrete crush resistance values*

Article 5 specified the possibility of increasing the mechanical resistance of concrete in compression by the installation of transverse reinforcement hoops (confinement of the concrete in the sense of Eurocode 2). The increase in resistance was then limited to:

$$60/100 \times \text{crush resistance}$$

The French Commission of Reinforced Cement assessed the coefficient of improved resistance as follows:

$$1 + m'V'/V$$

with:

– V′: volume of transverse or oblique reinforcements;

– V: volume of concrete for the same prism length;

– m′: coefficient of efficiency of the connections established with the longitudinal bars (values between 8 and 15, reaching between 15 and 32 for spiral transverse reinforcement).

Article 6 gave the value of the shear resistance of the concrete; this value was limited to:

$$10/100 \times \text{fatigue limit resistance} = 0.028 \times \text{crush resistance}$$

3.3.1.2. *Elastic strain*

The modulus of elasticity of the concrete was calculated from the characteristic diagram of the concrete (stress–strain diagram). By default, it was taken equal to 1/10 of that of steel, or approximately *Eb = 20,000 MPa*.

The steel/concrete equivalence coefficient, referred to as the density $m = E_a/E_b$, could vary from 8 to 15 and could increase if the breakage came from the concrete or decrease if it was due to an excess steel load.

3.3.2. *Reinforced concrete steels*

3.3.2.1. *Properties*

The steels commonly used in reinforced concrete constructions had a breaking strength of approximately 40–50 kg/mm^2. These values corresponded to structural steels whose elastic limits were approximately 22–24 kg/mm^2.

Article 17 also discusses the possibility of using:

– round bars (smooth rounds);

– profiled steel (H or I).

3.3.2.2. *Resistance*

The tensile stress of the steel was limited to:

$$0.5 \times \text{elastic limit of steel}$$

For parts subjected to alternating stresses (fatigue in the current sense), such as bridge slabs, the tensile stress was limited to:

$$0.4 \times \text{elastic limit of steel}$$

This value could be further reduced depending on the fatigue criteria selected.

Considering the elastic limits of section 3.3.2.1, the tensile strength values of the reinforced concrete steels were approximately *10 and 12 kg/mm^2*.

NOTE.– FeE22-grade steels were recognizable by the presence of two longitudinal splines. Experience shows that they were the cause of sudden breaks.

3.3.2.3. *Steel/concrete adhesion*

Article 6 of the directive fixed the value of the "longitudinal sliding on itself and its adhesion to the metal" at:

$0.10 \times R_b$, where R_b is the "fatigue" limit in the compression of concrete.

Thus, the values are identical to those obtained for the shear strength of concrete.

Mr. Bauchinger, a professor at the Technical University of Munich, attributed friction surface values between 40 and 45 kg/cm² to "tangential adhesion". The Reinforced Cement Commission considered this value to be too high and reduced it to 15–30 kg/cm². This difference was linked (according to Féret's experimental results) to the following factors:

– iron oxidation;

– nature of the cement (the CPAs adhered better than the slag cements);

– compressive strength of concrete (C/E ratio, granulometry, compactness, etc.).

Consequently, the continuity of overlapping bars was ensured using a common length of approximately *67 Φ* (Φ being the diameter of the bars). However, this value was valid only under certain conditions, and the overlap length l_r was then calculated as follows:

1) the steel/concrete adhesion was given by $S_f \times 0.10 \times R_b = \pi \times \Phi \times lr \times 0.10 \times R_b$;

2) the tensile force in the bar was $\pi \times \Phi^2/4 \times R_a$;

3) the equality of the two values allowed us to obtain $lr = 2.5 \times R_a/R_b \times \Phi$.

For concrete containing 350 kg of cement capable of supporting a compressive stress of 0.28 × 180 = 50 kg/cm² at 90 days and steel that can withstand a tensile stress of 12 kg/mm², we obtained a bar overlap length of:

$$Lr = 2.5 \times 12/0.5 \times \Phi = \mathbf{60\ \Phi}$$

NOTE.– These overlap lengths were relatively large and sometimes difficult to implement for straight lengths. It later became customary to provide hooks at the ends of the two facing bars. There was also the possibility of adding transverse reinforcements to the overlap area. These provisions were developed in the 1934 regulations.

3.4. Durability and cover of reinforcements

3.4.1. *Overview*

The durability criteria are not mentioned in the directive, nor are the exposure conditions of the works to physicochemical attacks.

3.4.2. *Durability requirement*

Service lifetimes are not mentioned.

3.4.3. *Cover*

Article 16 of the directive gives cover values of *15–20 mm*:

> The distances between the reinforcements and the walls of the formworks will be such that they allow the perfect compaction of the concrete and its tightening against the reinforcements. These latter distances, even when only mortar is used without gravel or stones, must always be at least 15 to 20 mm to protect the reinforcements from bad weather.

NOTE.– If we refer to the tests carried out in the 1990s by Mr. Balayssac, the carbonation depth values as a function of the mechanical resistance of the concrete are those indicated in Figure 3.1.

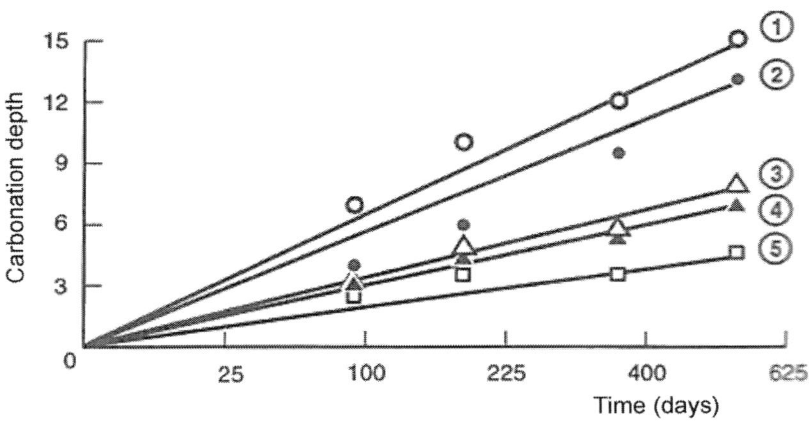

Figure 3.1. *Evolution of the carbonation depth as a function of time (Balayssac 1992). Curves 1–5: CPJ-CEM II 32,5 concrete with f_{c28} values of 20, 25, 30, 35 and 40 MPa, respectively*

COMMENT ON FIGURE 3.1.– *For concrete at 20 MPa, curve 1 indicates a carbonation depth of 15 mm, that is, the value of the cover, after approximately two years of exposure (under experimental conditions).*

3.5. Structural analysis

3.5.1. *Overview*

The purpose of structural analysis is to determine the distribution of stresses or constraints and strains. The behavior model used in the directive is that of *linear elastic behavior*.

Article 10 states that:

> Strength calculations will be made using scientific methods based on experimental data and not by empirical processes. They will be deduced either from the principles of the strength of materials or from principles that offer the same guarantee of accuracy.

3.5.2. *Loads cases and combinations*

Article 9 of the directive requested that resistance calculations should be considered:

– the greatest external forces;

– wind;

– snow;

– thermal forces (for non-freely expandable structures);

– effects of shrinkage (for non-freely expandable structures).

These loads were combined without load factors. However, the different types of combinations of actions were not defined.

3.5.3. *Second-order effects*

Second-order effects that can significantly affect the *overall stability* of the structure were not considered in the directive.

3.5.4. *Modeling the structure*

At that time, there were no digital means of modeling structures.

3.6. Calculating the structures

3.6.1. *Simple compression*

Simple compression was calculated from the fictitious homogeneous section Ω. The value of Ω is given by:

$$\Omega = \Omega_b + m\,\Omega_a$$

with:

- **Ωb**: concrete cross-sectional area;
- **Ωa**: total cross-sectional area of the longitudinal reinforcements;
- **m**: steel/concrete equivalence coefficient.

The calculation of the compressive stress on the concrete was equal to:

$$R_b = N/\Omega$$

The constraint on steel is as follows:

$$R_a = m\,N/\Omega$$

The coefficient m varied between 8 and 15.

NOTE.– The text warned against the systematic consideration of a coefficient m = 15 and proposed the following provisions:

– m = 8 when the diameter of the longitudinal bars was equal to 1/10 of the smallest dimension of the element with transverse reinforcements spaced at a maximum of this value;

– m = 15 when the diameter of the longitudinal bars was equal to 1/20 of the smallest dimension of the element with transverse reinforcements spaced at a maximum of one-third of this value.

The value m = 15 risked, according to the text, giving the metal a greater degree of resistance than it should normally have.

3.6.2. *Compression with bending*

The calculation was performed considering:

– the fictitious cross-sectional Ω (homogeneous section);

– the moment of inertia relative to the transverse axis of bending **I = Ib + m Ia**.

The constraints were then calculated using the classical formulas for material resistance, namely:

– *concrete compression on the extreme fibers*:

$R_b = N/\Omega + (M\ v_b)/I$ with v_b distance from the extreme fiber to the axis of inertia

The formula was modified to account for the slenderness through the radius of gyration r ($I = r^2\ \Omega$) according to:

$$R_b = N/\Omega\ (1 +/- (v_0 v_b/r^2))$$

with $\mathbf{v_0}$: eccentricity of N relative to the geometric axis;

– *compression of steel on the extreme fibers*:

$\mathbf{R_a = m\ (N/\Omega + (M\ v_a)/I)}$ with v_a distance from the extreme steel to the axis of inertia

– *concrete traction*: *Article 11* specified that the tensile strength of concrete (extension strength) was to be considered as zero in the calculation of the section strength. However, it could be considered in strain calculations;

– *steel tension*: see the following section on bending.

3.6.3. *Bending*

– *Simple bending*: it is calculated from the equilibrium equations of elastic forces. Where:

- the sum of the forces is (N = 0);

- the sum of the moments is equal to the applied bending moment M.

Consider the following equations:

$$0 = b'\ y_1^2/2 + (b - b')\ \varepsilon\ (y_1 - \varepsilon/2) + m\ \omega\ (y_1 - d) \quad [3.1]$$

$$M/K = b' y_1^3/6 + m\ \omega\ (y_1 - d)\ d - m\ \omega'\ (h - d' - y_1)\ (h - d') \quad [3.2]$$

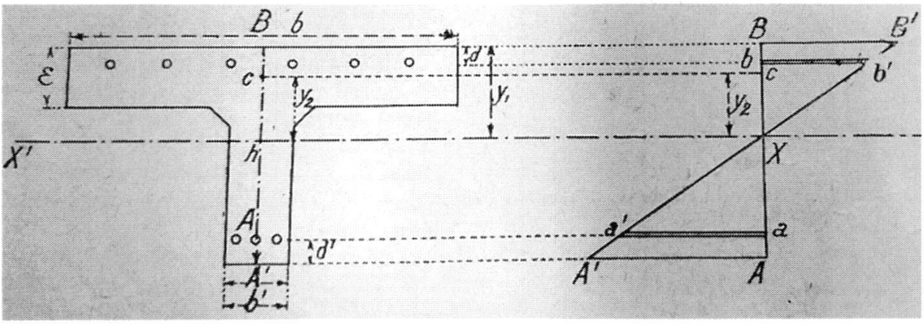

Figure 3.2. *Diagram of the bending calculation*

K is the angular coefficient.

Using these equations, the compressive stresses of the concrete and steel and the tensile stresses of the steel were calculated according to the following formulas:

$$R_b = Ky_1$$

$$R_a = m\,K\,(y_1 - d)$$

$$R'_a = m\,K\,(h - d' - y_1)$$

– *Simple compound bending*: the calculation was identical to that of simple bending, taking into account compression or traction N in equation [3.1] and considering that the moment of the elastic forces with respect to C is zero.

3.6.4. *Buckling*

Article 12 of the directive validated the fact that the parts whose slenderness was less than *20* and whose concrete compressive stress remained below the limit value were not checked for buckling. Otherwise, it was customary to use the Rankine formula:

$$N/\Omega\,(1 + kl^2/10\,000r^2) \leq Rb$$

with:

– N: total compressive force;

– l: length of the element;

– r: radius of gyration;

– k: coefficient depending on the fixity conditions at the ends according to Table 3.2.

In the case of a part undergoing composite bending (bending M + compression N), the Rankine formula was replaced by:

$$N/\Omega\ (1 + kl^2/10{,}000r^2) + M\ v/I \leq Rb$$

or:

$$N/\Omega\ (1 + kl^2/10{,}000r^2 + v_0 v/r^2) \leq Rb$$

Conditions at the ends	Value of k	Observations
An embedded support The other free	4	–
An embedded support The other articulated	0.5	In case of imperfect integration $0.5 \leq k \leq 1$
Two articulated supports	1	–
Two embedded supports	0.25	In the event of imperfect embedding of one of the supports: $0.5 \leq k \leq 1$ In case of imperfect embedding of the two supports: $0.25 \leq k \leq 1$

Table 3.2. *Coefficient k values according to the support conditions*

3.6.5. *Design*

3.6.5.1. *Simple compression*

The data are as follows:

– rectangular column 400 × 600;

– permanent loads G = 1.5 MN;

– variable loads Q = 0.5 MN;

– concrete quality: mechanical resistance of 20 MPa to crushing at 90 days, amount of cement 400 kg/m^3;

– buckling length $l = 4$ m;

– Fe 24-grade steel, 8 cm^2 steel section.

The total compression force on the column is as follows:

$$N = G + Q = 2 \text{ MN}$$

The compressive stress supported by the concrete is as follows:

$$R_b = N/\Omega = 2/(0.4 \times 0.6) = 8.33 \text{ MPa}$$

The compressive stress supported by the steel is as follows:

$$R_a = m\, N/\Omega = 10 \times 2/(0.4 \times 0.6) = 10 \times 8.33 = 83.3 \text{ MPa}$$

NOTE.– The value of m varies from 8 to 15 depending on the density of the longitudinal reinforcements, and buckling was not considered since $l/a \leq 1/20$.

These values are compared with the values of the admissible constraints mentioned previously, that is:

– 5.6 MPa for concrete;

– 96 MPa for steel.

The concrete cross-section is insufficient and would need to be multiplied by 1.5, that is, an overall cross-sectional area of 0.36 m^2 (instead of 0.24 m^2).

3.6.5.2. *Simple bending*

We consider a reinforced concrete slab with the following assumptions:

– span 1.50 m;

– thickness 0.08 m;

– five steel reinforcements of 8 mm per meter (As = 2.5 cm^2/LM);

– uniformly distributed variable load of 3 kN/m^2;

– concrete containing 300 kg/m^3 of cement.

The load per linear meter of slab is given by:

– self-weight of the slab, that is, $25 \times 0.08 = 2$ kN/m;

– variable load, that is, 3 kN/m.

This value is a total of 5 kN/m.

It was typical to consider that the slabs were partially embedded in their supports.

The span moment taken into account in the calculations was then given by:

$$M = pl^2/10$$

where $M = 5 \times 1.5^2/10 = 1.12$ kNm/m.

If we consider a 2-cm steel cover, the distance between the center of gravity of the steel and the upper face of the slab was: $a = 0.08 - (0.02 + 0.004) = 0.056$ m.

The lever arm of the elastic torque was then given by:

$$z = 8/9\ a$$

Here, $z = 0.05$ m.

The traction in the steel was equal to $F_t = 1.12/0.05 = 22.4$ kN. The constraint ("average fatigue" according to the meaning of the text) was equal to $2240/250 = 8.96\ \text{kN/mm}^2 \leq 12\ \text{kN/mm}^2$.

The compressive stress of the concrete was given by $2{,}240/(1.9 \times 100) = 11.80$ bar $= 1.2$ MPa ≤ 4.4 MPa (1.9 cm represents the distance between the neutral axis and the upper fiber of the slab).

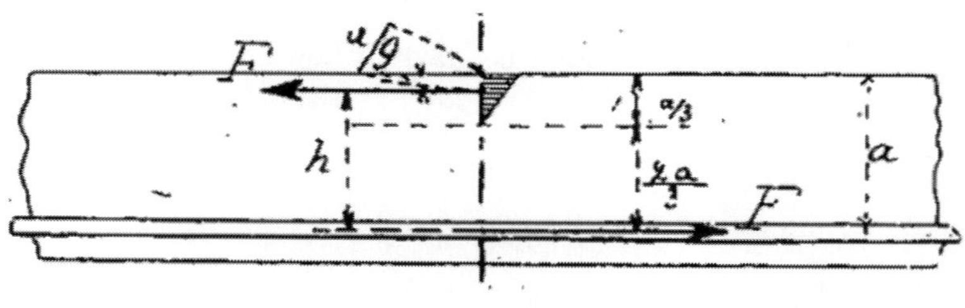

Figure 3.3. *Functional diagram of a beam*

NOTE.– The value of the bending moment at the mid-span $pl^2/10$ involved considering at the supports a negative bending moment whose value was generally fixed at $pl^2/40$.

This arrangement was permitted for floors supported by walls when the walls continued over them. In cases where the slabs were supported by walls without walls over them or by beams, the maximum moment to be considered at mid-span was equal to $pl^2/8$.

NOTE.– The case of slabs supported on four sides was also considered.

For example:

– solid slabs;

– or ribbed slabs, with the ribs being orthogonal.

In the case of a slab with span l in one direction and l′ in the perpendicular direction, the design method involved calculating the maximum bending moment in the direction of span l as with an isostatic slab bearing load in one direction only. The result obtained was multiplied by the coefficient:

$$1/(1 + 2 \times l^4/l'^4)$$

The same method was used in the l′ direction by swapping the letters l and l′ in the formula.

In the previous example, considering a square slab supported with the same degree of embedding at the supports, the value of the bending moment in the calculation was:

$$M_x = M_y = pl^2/10 \times 1/(1 + 2) = pl^2/30$$

This formula is consistent with that given in Appendix F3 of the BAEL regulations for the value of $\alpha = l_x/l_y = 1$.

The value of $\mathbf{M_x = M_y = 0.037\ pl^2}$. It should be noted, however, that this formula was chosen for a slab on simple supports without embedding (not even partial) at the supports.

3.7. Constructive provisions

The main provisions relating to the implementation of the structure were contained in Chapter 3 of the directive under the title “Erection works”.

3.7.1. *Pouring concrete*

It was recommended to use *slow setting cements* (S cement according to current European standards). The layers were "compacted" to obtain a final thickness of 5 cm for standard concrete, which was greater for gravel concrete.

3.7.2. *Distance between reinforcements and between reinforcements and formwork*

The distance between the steel as well as between the steel and the formwork had to allow:

– the "compaction" of the concrete;

– its tightening against the frames;

– a minimum cover of 15–20 mm to meet the durability criterion.

3.7.3. *Construction joints of concrete*

The construction joints were treated as follows:

– "intense" cleaning of old concrete;

– wetting of the existing concrete surface;

– pouring of fresh concrete.

3.7.4. *Curing of the concrete*

Article 2 of the directive recommended maintaining the concrete wet for a period of 15 days after pouring.

NOTE.– On July 11, 1912, following certain observations made by bridge and road chief engineers, the General Council of Bridges and Roads clarified the calculation methods concerning hollow reinforced concrete slabs.

1) Distribution of an isolated load on a rectangular hollow slab (a × b): At the center of the hollow slab (point 0), the bending moments are called M_a and M_b, the bending constraints R_a and R_b, p is a load uniformly distributed over the entire hollow slab and P is a point load at the center O of the hollow slab.

2) The bending moments are then given for a load distributed p by:

$M_a = pa^2 b^4/8(a^4 + a^2b^2 + b^4)$

$M_b = pb^2 a^4/8(a^4 + a^2b^2 + b^4)$

$F = pa^4b^4/80EI\ (a^4 + a^2b^2 + b^4)$

3) The bending moments are then given for a point load P by:

$Ms_a = P\ (a^2 + b^2)/(5a^2 + 3b^2)$

$M_b = P\ (a^2 + b^2)/(3a^2 + 5b^2)$

$f = Pa^2b^2/40EI\ (a^2 + b^2)$

I is the modulus of inertia of the cross-section.

4

The January 1913 Technical Note on the Calculation of Reinforced Concrete Works

4.1. Overview

This note covered reinforced concrete works in buildings constructed by the engineering sector that appeared in the memorandum of September 11, 1912, and also in the provisional note of December 5, 1903. It also appeared in the 1906 Technical Note main recommendations.

The modified requirements of the 1903 note became obligatory because reinforced floors whose calculation satisfied the strict application of the methods recommended by the said note.

NOTE.– It contains 25 paragraphs divided as follows:

– section I: compressed elements;

– section II: bent elements;

– section III: elements not falling into the previous categories.

4.2. Calculation bases

This new note kept the same calculation method (provisional 1903 note of and 1906 Technical Note). In all cases, the tensile strength of the concrete was not considered in the calculations.

4.3. Materials

4.3.1. *Concrete*

4.3.1.1. *Resistance*

For compressed elements (columns), the compressive stress of the concrete was limited to the following values:

– for concrete containing 300 kg of cement, 800 L of gravel and 400 L of sand: *44.8 bar* (4.48 MPa);

– for concrete containing 350 kg of cement, 800 L of gravel and 400 L of sand: *50.4 bar* (5.04 MPa).

For the bent elements (beams, slabs, etc.), the compressive stress of the concrete was limited to the following values:

– for concrete containing 300 kg of cement, 800 L of gravel and 400 L of sand: *22.4 bar* (2.24 MPa);

– for concrete containing 350 kg of cement, 800 L of gravel and 400 L of sand: *25.2 bar* (2.52 MPa).

These values of mechanical resistance in compression for bending corresponded to values of pure compression with a safety coefficient of 2.

4.3.1.2. *Elastic strain*

The elastic moduli of the steel and concrete were 200,000 MPa and 20,000 MPa, respectively. That is, $E/E' = 10$ (with E being the modulus of steel and E′ being the modulus of concrete). In the 1903 note, this value had been set equal to 20.

4.3.2. *Reinforced concrete steels*

4.3.2.1. *Properties*

For bent elements, iron or steel could be used. The average stress considered in the calculations was as follows:

– 9 kg/mm^2 for iron (90 MPa);

– 12 kg/mm^2 for steel (120 MPa).

4.3.2.2. *Steel/concrete adhesion*

The provisions of the 1906 note were validated.

4.4. Durability and cover of reinforcements

The recommendations of the 1906 note were considered.

4.5. Structural analysis

The recommendations of the 1906 note were considered.

4.6. Calculation of the structures

4.6.1. *Simple compression*

In the 1903 version, the posts ("pillars") were sized by equalizing the elastic shortenings of the concrete and steel under load. The forces were shared according to the ratio $E\ S_a/E'\ S_b$, where E_N is the elastic modulus of the steel, E′ is the elastic modulus of the concrete, S_a is the cross-section of the post reinforcements and S_b is the concrete cross-section.

It was accepted that $E = 20E'$ ($E = 200{,}000$ MPa and $E' = 10{,}000$ MPa; this value corresponds to the long-term BAEL modulus). The distribution ratio therefore became $20S_a/S_b$. The total compressive stress of the concrete (including steel) was then limited to 25 bar (2.5 MPa).

CALCULATION EXAMPLE.– Let us take a $20 \times 20\ \text{cm}^2$ square cross-section post reinforced with 4 bars with a diameter of 13 mm to support a load of 11,000 kg (it should be noted that in the calculations, the loads are always perfectly centered).

The compressive stress of concrete ("fatigue" according to the text) is given by:

$$11{,}000/(0.04 + 20 \times 0.000531) = 22\ \text{kg/cm}^2$$

With this value being less than 25 kg/cm², the post was correctly sized.

In the 1913 version, the posts ("pillars") were sized considering an admissible compressive stress for the concrete of 44.8 kg/cm² (instead of 25 kg/cm²).

Consequently, if we take the previous example, to take an identical effort of 11,000 kg, with the same reinforcements, a post of 14 cm by 14 cm was sufficient.

4.6.2. *Bending*

In the 1903 note, the calculation was carried out by considering:

– that the tensile strength of the concrete was ignored;

– that the compressive strength of the reinforcements was initially ignored;

– that the neutral axis of the heavily reinforced elements (1–2%) was located at d/2 (d being the distance between the center of gravity of the tensioned reinforcements and the upper surface of the element);

– that the neutral axis of the weakly reinforced elements was located at d/3 (upper third).

Under these conditions, the resistance values of the materials were as follows:

– 9 kg/mm² for iron (quality giving 32 kg of strength at break for an elongation of 8 %);

– 12 kg/mm² for steel (quality giving 42 kg of strength at break for an elongation of 22 %);

– 25 kg/cm² in compression for concrete.

The same provisional note of 1903 discussed shear force reinforcement.

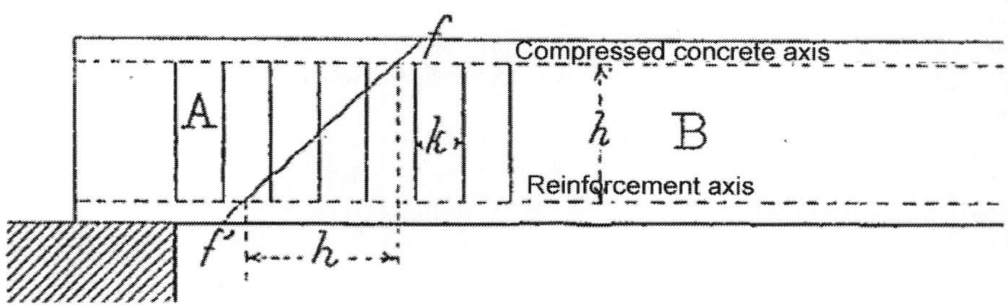

Figure 4.1. *Principle of shear force reinforcement*

The transverse reinforcements (called "stirrups" in the note) were calculated using the following formula:

$T/r \leq h/k\ S_t$

with:

– T: shear force value;

– r: tensile stress of the stirrups;

– k: average spacing of the stirrups in the support area;

– S_t: cross-section of a stirrup.

EXAMPLE.– In some beams, it was possible to find transverse reinforcements composed of flat iron strips of 0.02 m by 0.003 m. They were spaced approximately 10 cm apart in the support area.

In the 1913 version, the above-mentioned values were retained.

4.6.2.1. *Calculation example 1*

We consider a reinforced concrete slab with the following assumptions:

– span 1.50 m;

– thickness 0.08 m;

– five steel reinforcements of 7 mm per meter ($A_s = 1.92$ cm²/lm);

– uniformly distributed variable load of 3 kN/m².

The load per mL of slab is given by:

– self-weight of the slab is $25 \times 0.08 = 2$ kN/m;

– variable load is 3 kN/m.

That is, the total load is 5 kN/m².

The bending moment was $M = pl^2/10$ or 112 kgm.

The steel cover (slab located inside a building) was then considered to be approximately 1.5 cm.

The value of d was $d = 0.08 - (0.015 + 0.004) = 0.061$ m.

The lever arm was $z = 0.9d = 0.9 \times 0.061 = 0.055$ m.

The tensile force in the lower fiber and the compressive force in the upper fiber were therefore:

$$F = M/z = 112/0.055 = 2{,}036 \text{ kg}$$

The tensile stress of the steel was 2,036/1.92 = 1,060 kg/cm² (106 MPa) ≤ 1,200 kg/cm².

The average compressive stress of the concrete was 2,036/6.1/3 = 10 kg/cm² ≤ 25 kg/cm².

4.6.2.2. *Calculation example 2*

Consider the floor shown in Figure 4.2 on the formwork/reinforcement plans.

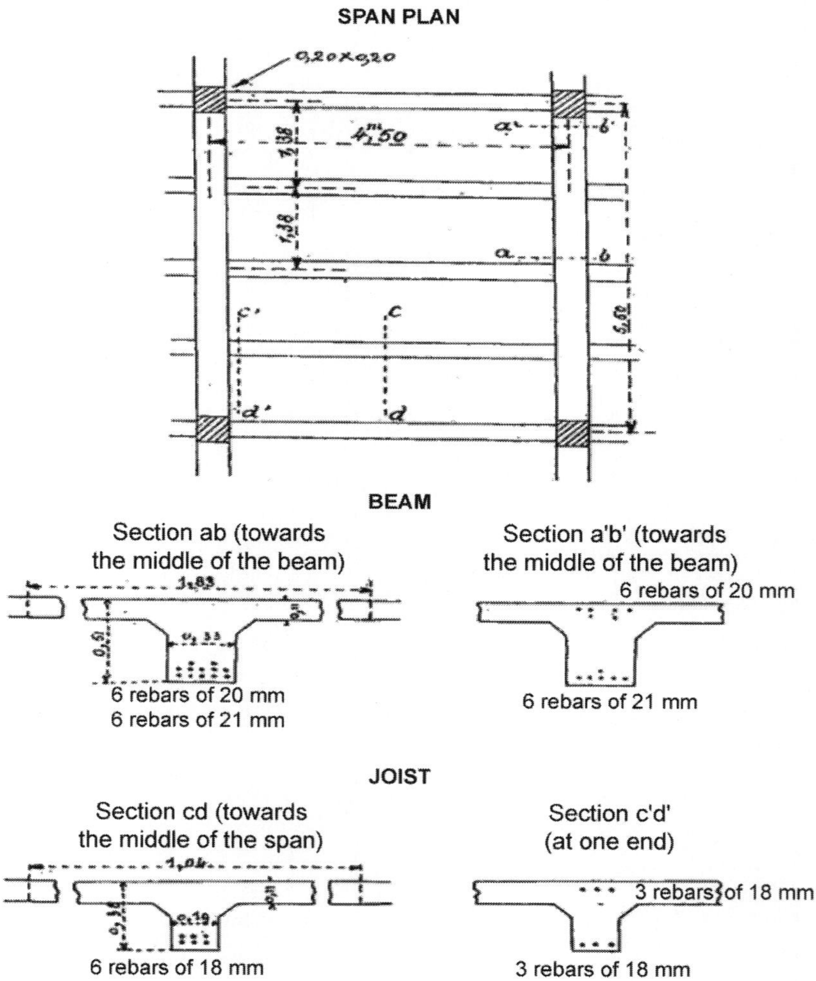

Figure 4.2. *Floor structure plans and cross-sections*

This is the floor of a military building designed to support a load of 950 kg/m². The concrete used contains 300 kg/m³ of cement.

The purpose is to check the main beam whose squaring is as follows:

– height 0.51 m;

– width 0.33 m;

– lower fiber reinforcements: six steels with a diameter of 20 mm (cross-section 18.84 cm²) and six steels with a diameter of 21 mm (cross-section 20.8 cm²).

Calculation of applied loads

– slab: 4.50 × 0.11 × 2,500 = 1,288 kg/ml;

– beam (including joists): 645 kg/ml;

– live load: 4.50 × 950 = 4,275 kg/ml.

Calculation of stresses

The mid-span bending moment was estimated at an intermediate value of $pl^2/10$.

The mid-span bending moment is: $M = 6{,}208 \times 5.50^2/10 = 18{,}780$ kgm.

NOTE.– In the regulation, the span was considered the center distance between the supports.

Calculation of forces in the beam

Considering a 3 cm cover of the lower steels, the value of d would be: 0.51–0.03–0.02 = 0.46 m.

The lever arm z was then estimated at 0.9d or 0.414 m.

This resulted in a tensile force in the lower fiber and a compressive force in the upper element of: F = 18,780/0.414 = 45,362 kg.

Checking the stresses

The tensile stress in the steel was therefore 45,362/(18.84 + 20.8) = 1,144 kg/cm² (114 MPa). This value was compared with that of 1,200 kg/cm² (120 MPa). The beam was therefore correctly reinforced.

The compressive stress of the concrete was determined by considering:

– the slab involved was equal to Min (l/3 l; 0.75 e), where l is the span of the beam and e is the center distance of the beams, that is, in our case, Min (5.50/3; 0.75 × 4.50) = 1.83 m;

– the compressed zone of the beam was located above d/3, or 0.15 m.

If we consider the slab only, a cross-section of 0.11 × 1.83 = 0.20 m², the value of the concrete compression is 45,362/2,013 = 22.5 kg/cm², whereas it is 22.4 kg/cm² for a concrete dosage of 300 kg. Overall, the dimensioning was acceptable.

Shear force verification

The verification was called verification under vertical shear force.

The shear force of the isostatic beam is: T = 13,300 kg. The shear stress is given by T/S × b = 13,300/(51 × 33) = 7.9 bar. The shear stress value was set at 8 bar. The shear stress of the concrete was therefore considered acceptable.

The section of the transverse steel stirrups was given by the formula k = ωRh/T, where k is the spacing between the stirrups, ω is the total cross-section of the six stirrups (counting the two pieces), and h is the lever arm of the internal forces. Therefore, h = 0.415 m, ω = 12 × 50 mm² = 600 mm².

The stirrups will be made of flat iron with a maximum shear stress of 6.4 kg/mm².

This results in a separation of the stirrups of approximately k = 600 × 6.4 × 0.415/13,300 = 0.12 m.

NOTE.– Some steel diameters do not correspond to those commonly used today.

4.7. Examples of finished structures

4.7.1. *1927 tank*

This is a tank structure whose formwork plans are given in Figures 4.3–4.6.

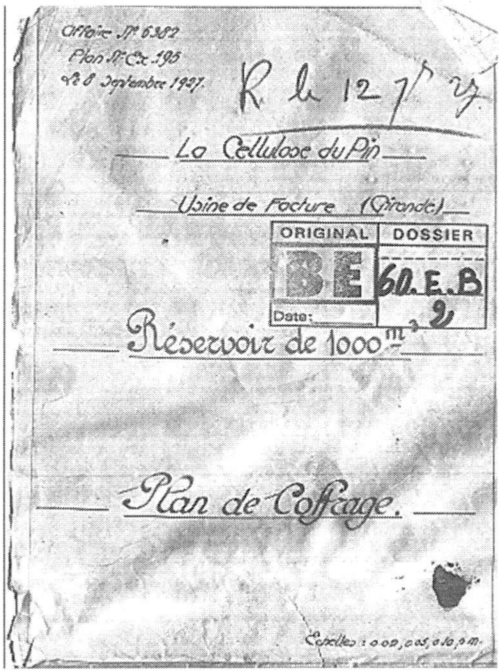

Figure 4.3. *Cover page of the formwork plan (in French)*

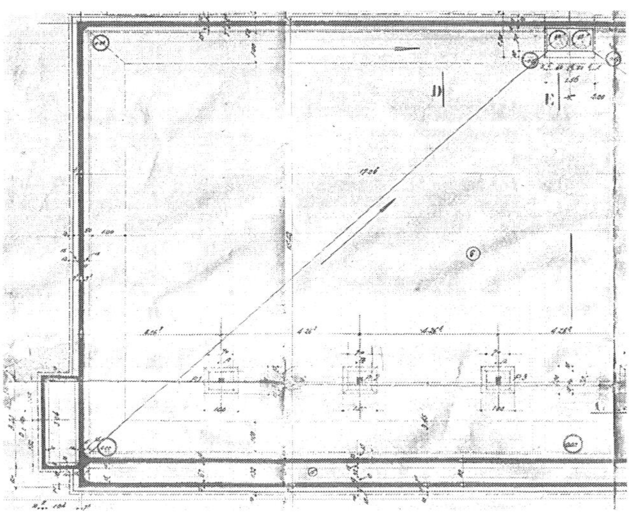

Figure 4.4. *Plan view (1/2 basin)*

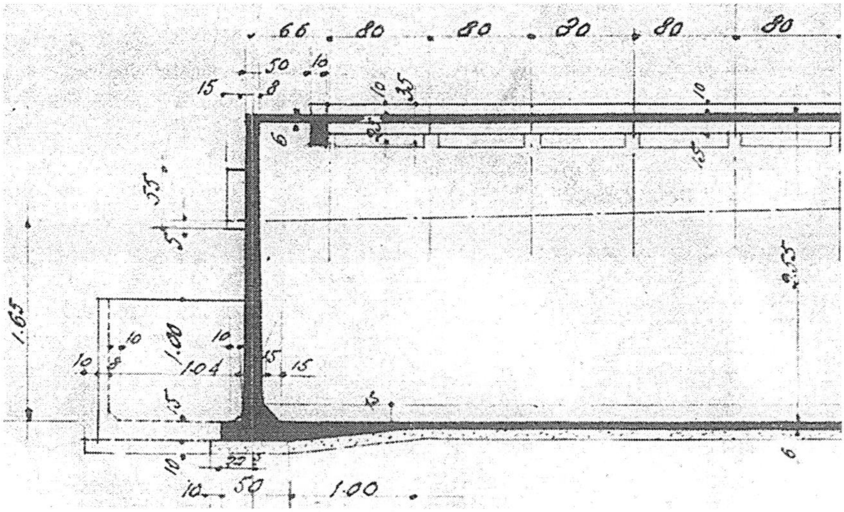

Figure 4.5. *Cross-section of the field element*

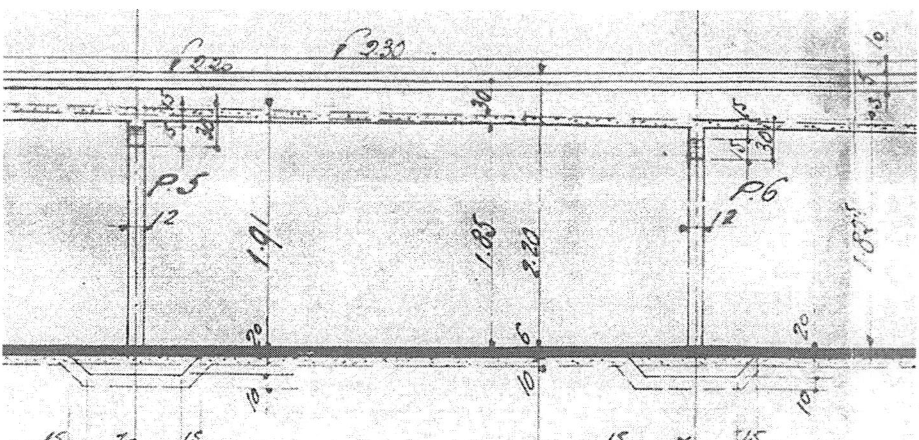

Figure 4.6. *Cross-section of the interior posts and gutter*

Views of the structure taken in 2020 during a periodic inspection are shown in Figure 4.7.

Figure 4.7. *Views inside the tank and of the tank walls during an inspection in 2022. For a color version of this figure, see www.iste.co.uk/lauzin/concrete.zip*

Pachometer tests, used to estimate the thickness of the steel cover, carried out on the eastern wall indicate a thickness between 2.7 cm and 4.5 cm (depending on the layers).

The 2.7 cm cover likely corresponds to the first layer of reinforcement on the outer side of the wall, which, for a thickness of 10 cm, would correspond to a 2.5 cm cover for the steel layer located on the liquid side.

Sclerometer tests, aimed at estimating the apparent resistance of the concrete, carried out on the eastern wall indicate an average Rc value of approximately 36.7 MPa, which is satisfactory given the intended use of the structure and the durability of the material.

Notably, the squarings are very weak:

– thickness of wall 2.55 m in height: 8 cm at the top and 15 cm at the bottom of the wall (excluding gusset);

– thickness of the raft: 15 cm under the walls and 6 cm in the center of the tank.

The construction provisions consist of a wall/base embedding a gusset as well as a wall/wall embedding gusset in the corners of the tank.

Figure 4.8. *Wall/base embedding gusset. For a color version of this figure, see www.iste.co.uk/lauzin/concrete.zip*

Defects	Cause	Solutions
Construction defects		
Penetrating cracks affecting all vertical walls	Concrete shrinkage and failure to take thermal gradient into account	Repair of cracks from within the tank and sealing/waterproofing of the inside of the tank
Degradation of anchors of the additional prestressing cables	Degradation of the protection of the prestressing cables and corrosion of the cables and anchor heads	Study and implementation of a new structural reinforcement system
Lack of waterproofing and protection of concrete	The waterproofing of the tank is old and no longer protects the concrete	A chemical analysis of the effluent in accordance with FDP 18.011 Internal sealing/impermeabilization of the tank
Exposing the steel (inside the tank 1)	Insufficient cover, degradation linked to the degradation of the protective interior cover and to chemical attack of the effluent	A chemical analysis of the effluent in accordance with FDP 18.011 Internal sealing/impermeabilization of the tank
Leaching of concrete in the upper element of the cover, on tank 1	Degradation of the protective interior cover and chemical attack on the concrete by the effluent	A chemical analysis of the effluent in accordance with FDP 18.011 Internal sealing/impermeabilization of the tank

Table 4.1. *Main defects noted during the inspection*

It is also worth noting the recent implementation of an additional pre-stressing system intended to prevent the corners from opening.

The main structural defects are summarized in Table 4.1.

NOTE.– In 1927, there was a change in the regulations of loads and their calculation: in the calculation of the compressive and tensile stress limit of steels, the value of 0.4 fe in the event of impact was removed.

NOTE.– In Charles Rabut's 1910 course, the specifics of tank walls were discussed:

Tank walls must not have the same shape as the retaining walls. If the earth load can be used to balance the thrust, it is because the weight only acts on the upper surface of the berm... The greatest difficulty is to seal the cover. This is achieved when the load is low using a cover of rich cement mortar. If it is done afterwards, this cover must be very thin, otherwise it would undergo considerable shrinkage, no longer being prevented by the adhesion with the concrete block which would have already set when the cover was applied.

Under an average water load, the concrete is not waterproof, but it becomes so in the long run. When the structure is not urgently required, it is possible to wait a few months for complete waterproofing.

In the case of heavy loads, there is only one sure solution: continuous metal cover... To protect it from rust, it must be covered with a certain thickness of concrete.

4.7.2. *1928 tanks*

J. Gilbert and E. Mondon's 1828 *Traité d'adductions et de distributions d'eau* provides interesting information on the calculation and construction of concrete tank from this period. Section III "Overview" specifies:

> Reinforced concrete lends itself well to the creation of the various shapes required by water tanks. This construction method involves embedding a metal frame in a thin cement mortar.
>
> The reinforcement is most often made up of steel bars, round, square, flat or profiled in T, I, +, L, C… which are connected to form a mesh.

In accordance with the ministerial note of October 20, 1906, the elements of the metal frame must be calculated so that it can withstand alone the forces that tend to break it. The concrete only ensured the waterproofing and protection of the metal against oxidation.

4.7.2.1. *Example of a 170 m³ water tank*

The underground tank offered the most diverse solutions. It is generally composed of a circular structure resting on a foundation raft. The walls were thus often considered embedded at their base. The tank was covered by a dome. The foundation consisted of a layer of coarse concrete varying in thickness on which a thin layer of reinforced concrete was poured. A waterproofing screed was then added.

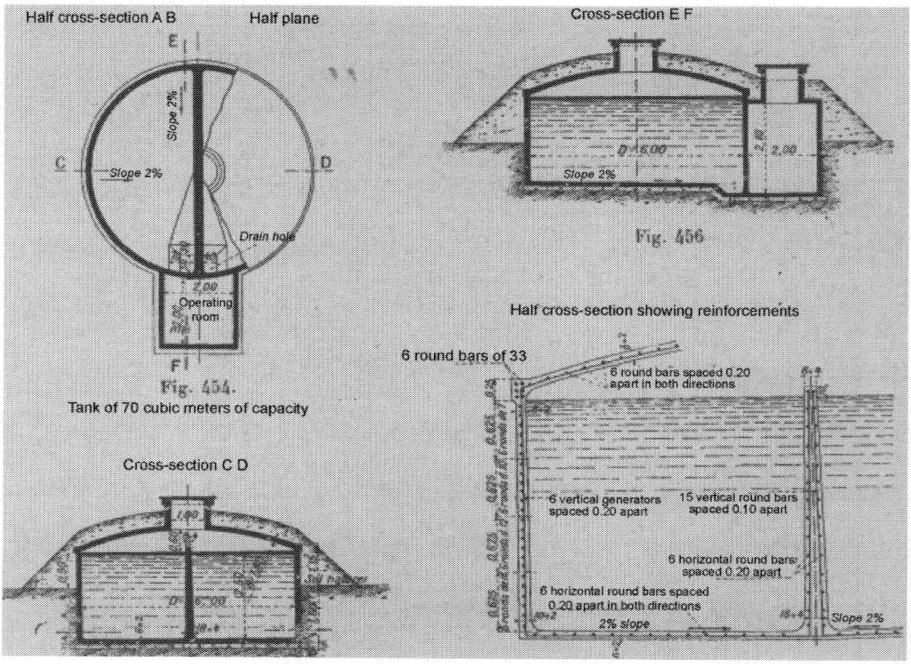

Figure 4.9. *Example of a 170 m³ buried tank. For a color version of this figure, see www.iste.co.uk/lauzin/concrete.zip*

The structural calculations were carried out according to the 1906 guidelines with the following assumptions:

– concrete: containing 300 kg/m^3 cement, 0.4 m^3 sand and 0.8 m^3 gravel;

– concrete working rate:

- in compression: 44.8 bar,

- in shear: 4.48 bar,

- in tension: 0 bar;

– steel: mild steel with an elastic limit of 24 kg/mm^2;

– steel working rate:

- in tension and compression: 12 kg/mm^2,

- in shear: 9.6 kg/mm^2.

The squarings selected for the project were:

– at the covering dome: diameter 9.40 m, spire 1.17 m, thickness 9 cm;

– at the walls: height 2.50 m, internal diameter 9.34 m and thickness 7 cm (internal water height 2.50 m);

– at the raft level: thickness 7 cm.

The structure was verified according to the 1906 guidelines as given in the following sections.

4.7.2.1.1. Checking the covering dome

The verification carried out with the formulas from section 2.5 leads to a compressive stress of 9.7 bar in the concrete of the arch for an acceptable stress of 44.8 bar. The construction provisions led to the implementation of a mesh composed of 7 mm bars spaced 20 cm apart for the hoop lines and 8 mm bars spaced 20 cm apart for the meridian lines. The lower belt of the covering dome bears the horizontal thrust (q = Mt of bending/deflection), that is, a traction of 36.8 t exerted by the following steels:

– 5Φ26;

– 1Φ22;

– 1Φ9.

It is for a total cross-section of 31 cm².

The working rate of the steels was 36,800/3,100 = 11.87 kg/mm² $\leq$ 12 kg/mm².

4.7.2.1.2. Verifying the wall

The thrust per linear meter of wall of a cylindrical tank was given by $Q = 1{,}000\ H^2/2$. The traction in the thickness of the wall then becomes $T = QD/2 = QR$.

The total cross-section of the hoops, for a steel working rate of 12 kg/mm², was therefore $250 \times 2.50^2 \times 9.34/12 = 1{,}220$ mm².

4.7.2.1.3. Verifying the raft

The total weight of the tank structure, the water contained and the weight of earth over the covering dome was 5.841 t/m².

The stress on the ground at the level of the base under the cover was considered uniform and expressed as:

$$5{,}841/(40 \times 100) = 1.50 \text{ bar}$$

This value was compared to the admissible stress of the soil in place.

J. Gilbert and E. Mondon specified that the soil supporting the tank had to be "solid and impermeable". The preferred solution therefore consisted of "a partially or totally buried circular tank, resting on the ground with a vertical wall widened at the base. This increase in width, which is generally done both outside and inside, serves to distribute the loads due to the covering if there is one and ensures that the vertical wall is embedded. Its thickness can vary by decreasing from 0.20 to 0.08 m from bottom to top".

NOTE.– Similarly, it was also possible to produce larger capacity structures, such as the one described below, comprising two adjoining tanks of 750 m³ each.

The tanks were cylinders 17.90 m in diameter with a water height of 3 m. They had a dome covering with a diameter of 7.90 m and height of 1 m. The dome rested elemently on reinforced concrete columns 30 cm in diameter and 3 m high, and the tanks rested on a concrete base of gravel and lime mortar 30 cm thick. To the right of the columns, an extra thickness of 60 cm was planned on an 80 cm crown.

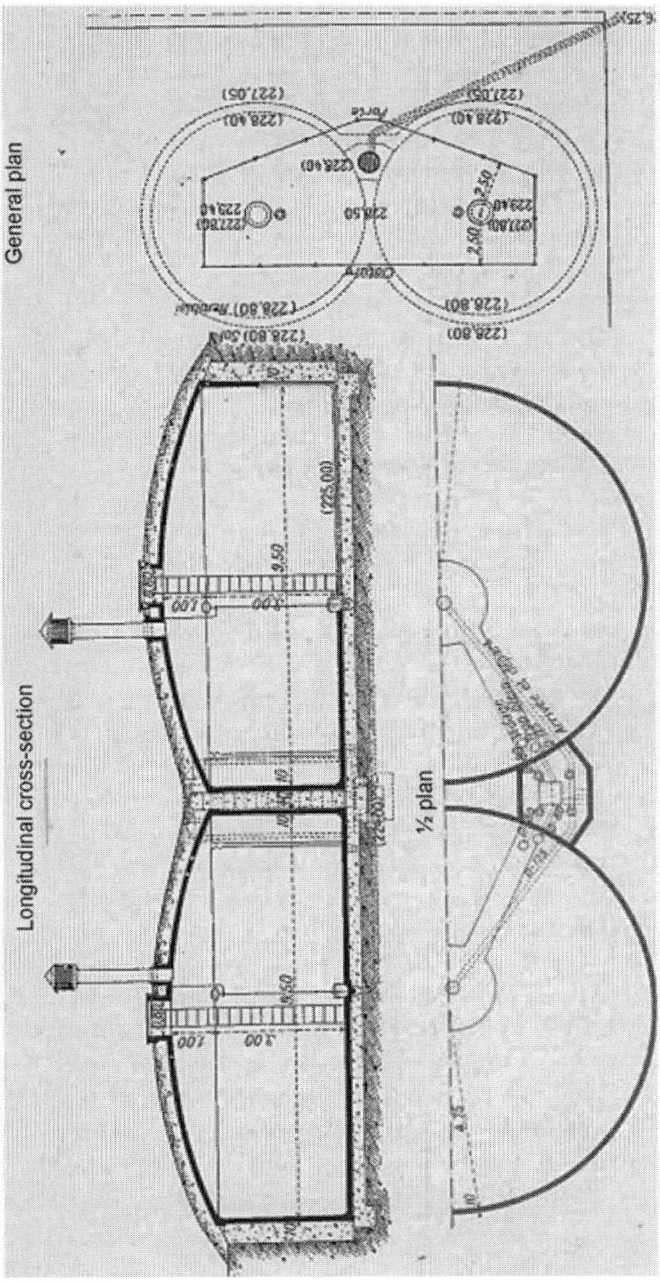

Figure 4.10. *Longitudinal cross-section of a tank with two adjoining walls. For a color version of this figure, see www.iste.co.uk/lauzin/concrete.zip*

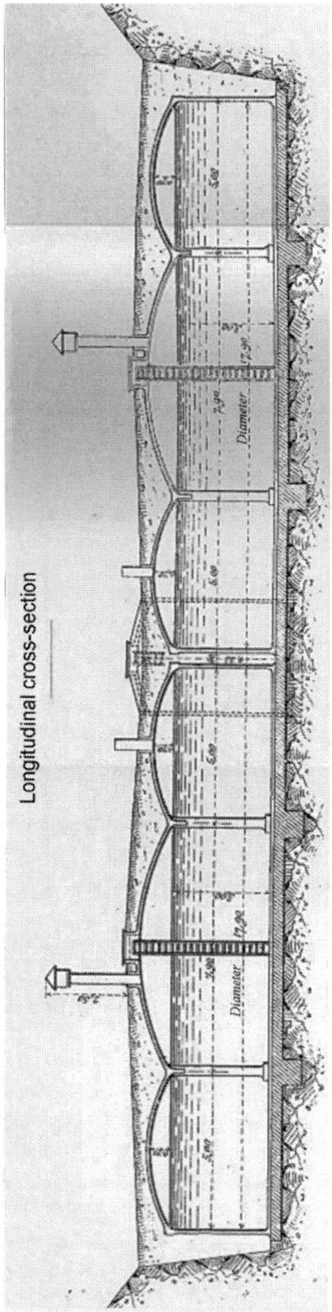

Figure 4.11. *Longitudinal cross-section of the tank. For a color version of this figure, see www.iste.co.uk/lauzin/concrete.zip*

4.7.2.2. *Example of a tower tank of 60 m^3*

The work presented here involves a tower tank studied by the Piketty company whose characteristics are as follows:

– a 60 m^3 tank surrounded *by a double wall to avoid temperature variations in accordance with the July 12, 1924 guidelines*;

– a valve chamber consisting of three raised elementitions between the posts supporting the tank. A ceiling closes the space.

Between the inner wall and the outer wall, it was common to add thermal insulation such as cork or sawdust.

NOTE.– Following the collapse of an oil tank supported by four pylons in 1928, doubts arose regarding the calculation of the pylons supporting the above-ground tanks. For a system of water towers on several pylons, it was assumed that the structure consisted of a platform (reinforced slab) whose dimensions were small compared with the height of the pylons, which allowed it to be considered rigid.

To study such a system, it was necessary to know the support reactions. Since statics does not allow these elements to be calculated from the moment when the number of supports exceeds 3, it was then necessary to consider deformable pylons.

In an article from 1928, M. Callandreau proposed the following method based on the previous assumptions.

In the common case of a regular polygonal plate within a circle with a radius r, if we considered that the first support was located on the Ox axis, then the coordinates of the points of application of the support reactions were given by:

$$x_k = r \cos 2k\pi/n \quad y_k = r \sin 2k\pi/n$$

The support reactions were as follows:

$$P_k = R\,[(r \cos 2k\pi/n)/(\Sigma\, x_k^2) + (r \sin 2k\pi/n)/(\Sigma\, y_k^2) + 1/n]$$

with:

– R: resulting from vertical loads;

– N: number of pylons.

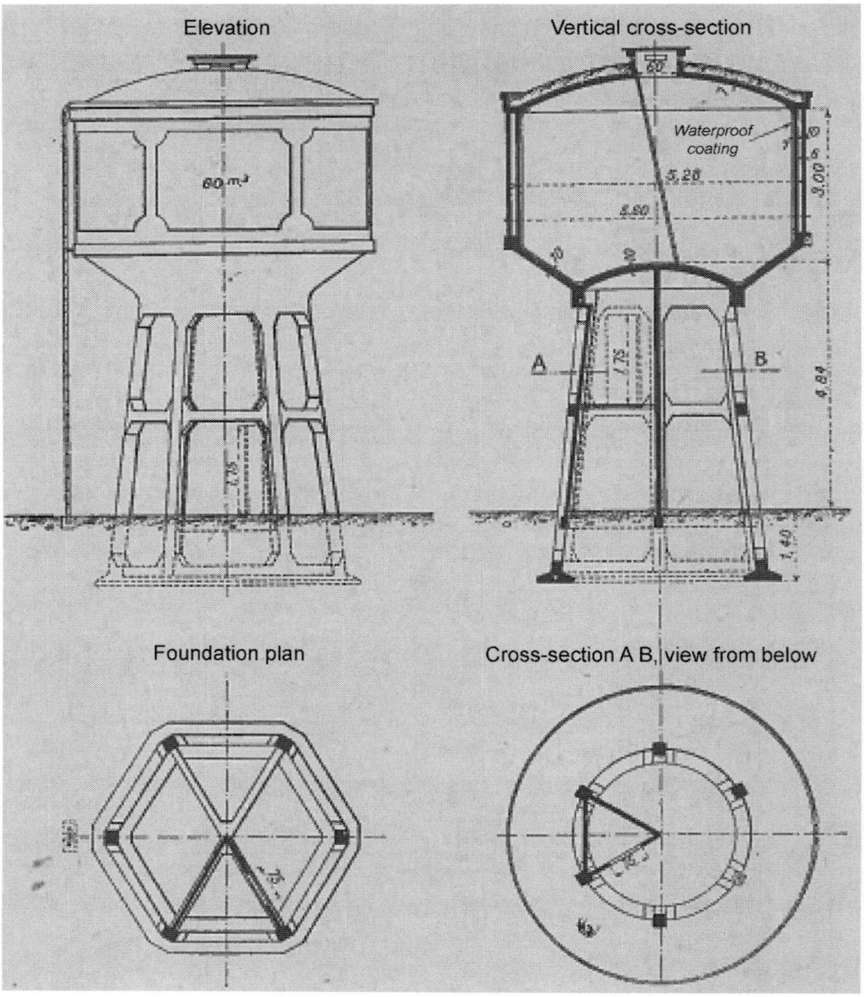

Figure 4.12. *60 m³ Piketty tower tank. For a color version of this figure, see www.iste.co.uk/lauzin/concrete.zip*

4.7.2.3. *Example of a tower tank of 330 m³*

The structure is a drinking water distribution tank. It is built on a 14.98 m high shaft and includes two compartments:

– central compartment with a capacity of 250 m³;

– a peripheral compartment of 80 m³.

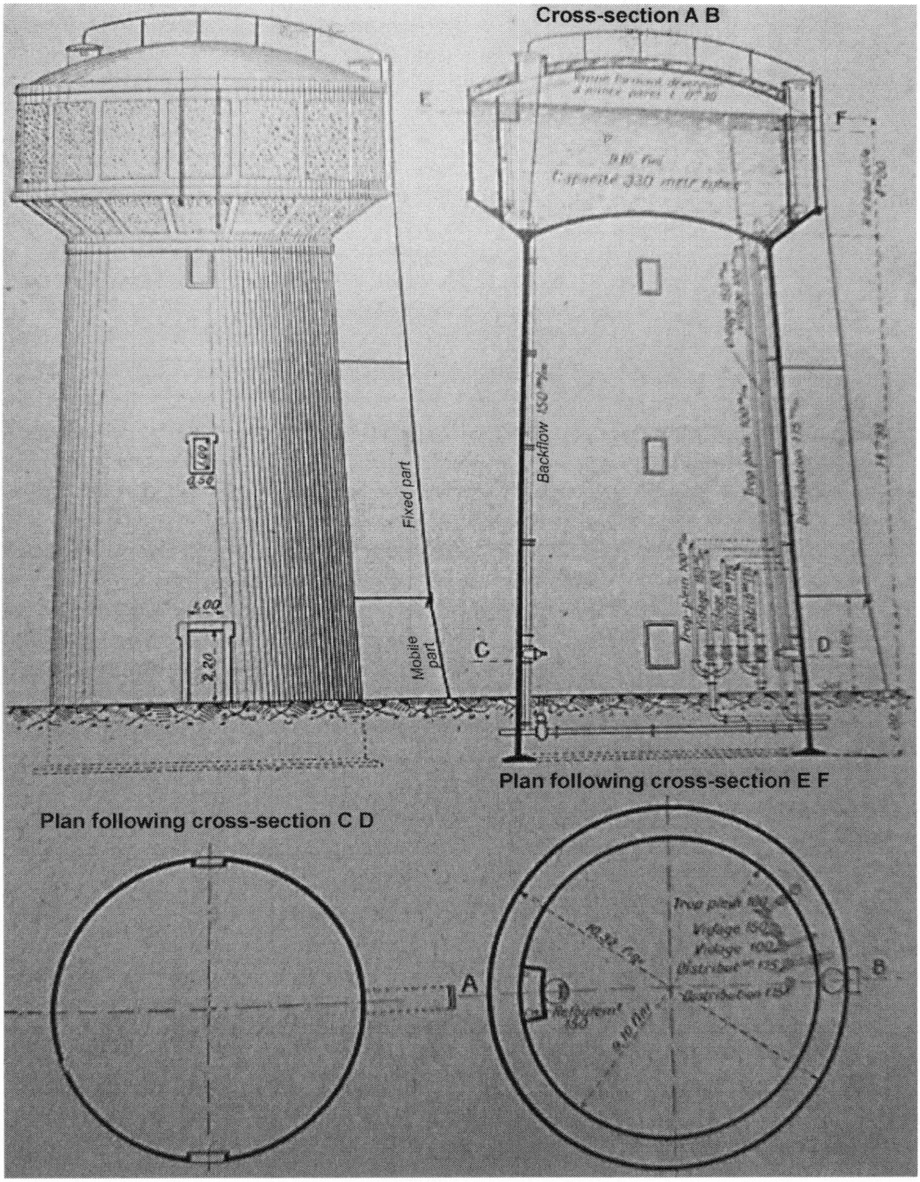

Figure 4.13. *Tower tank of 330 m³: cross-section and general view*

In winter and mild weather, the 1924 recommendations were that the two compartments could be used for drinking water. In very hot or very cold weather, the peripheral compartment, which was previously filled, was used to isolate the central compartment, which was the only one used for drinking water. The water in the peripheral compartment was then used for protection against fire.

The tank was ventilated by vents in the manholes.

To prevent the temperature in the tank from rising, 30 cm of earth covered a waterproof cement cover under the dome cover. The exterior elements of the structure were then coated with a cement mortar, and the hydraulic walls were coated with a 2-cm thick cement mortar to ensure waterproofing.

The assumptions when sizing the reinforced concrete structures were as follows:

– dome cover: 1/8 lowered dome, with a diameter of 11.03 m, a 1.38 m spire and a thickness of 7 cm;

– peripheral tank walls: wall height 3.58 m, water height 3.08 m, average internal diameter 11.03 m and thickness 7 cm;

– walls of the central tank: wall height 3.81 m, water height 3.81 m, average internal diameter 9.21 m and thickness 7 cm;

– connecting cone: the diameter varies from 11.03 to 7.72 m, and the thickness is 13 cm;

– bottom dome: dome lowered to 1/8, with a diameter of 7.72 m, a 0.97 m spire and a thickness varying from 13 cm at the base to 10 cm at the top;

– foundation slab: 1.05 m wide by 0.20 m high. The pressure on the ground was limited to 3 bar.

The structure was checked according to the 1906 guidelines as follows.

4.7.2.3.1. Checking the dome cover

The load distributed per m^2 of dome was given by:

– self-weight: $0.07 \times 2{,}500 = 175$ kg/m^2;

– watertight screed: $0.02 \times 2{,}000 = 40$ kg/m^2;

– soil: $0.30 \times 1{,}600 = 480$ kg/m^2;

– snow: 50 kg/m^2.

That is, the total charge is $P = 755 \times \Pi \times (f^2 + d^2/4) = 77$ t.

The vertical load per ml at the level of the lower belt of the dome was: $p = P/(\Pi d) = 2.22$ t/ml.

The lowered arch generated horizontal thrusts q = bending moment/deflection = $2P/(\Pi d) = 4.44$ t/ml.

The compression force was therefore $N = (2.22^2 + 4.44^2)^{1/2} = 4.97$ t/lm.

Hence, the compressive stress was $4{,}970/(100 \times 7) = 7.10$ bar.

This value was compared with the admissible compressive stress of the concrete, which was 44.8 bar for a concrete containing 300 kg/m^3 of cement.

The minimum reinforcement used was generally composed of:

– mild steel Φ7 every 20 cm in circles;

– mild steel Φ8 every 20 cm in meridians.

At the level of the lower belt of the covering dome, supporting the arch thrust $T = q\, d/2 = 24.3$ t/ml required a steel support measuring $24{,}300/12 = 2{,}025$ mm^2.

The reinforcement used was generally composed of 4Φ26 (2,123 mm^2).

4.7.2.3.2. Checking the walls of the peripheral tank

The thrust of the water on the walls of the tank was calculated by $Q = \rho H^2/2 = 1{,}000\ H^2/2$.

The traction generated in the wall was then estimated at $T = QR = 250H^2D$.

The steel section required to take up this tensile force was as follows: A_s = T/steel work rate.

Therefore, $A_s = 250 \times 3.08^2 \times 11.03/12 = 2{,}170$ mm^2.

These hoops were then distributed to take up the applied tensile force on each horizontal section of 1 m.

NOTE.– The tensile force in the wall could be calculated from the total diameter of the tank but also by taking into account the tank partition.

The lower belt of the tank was sized to absorb the forces by:

– the load of the cylindrical wall comprising:

- the weight of the covering dome, that is, 77 t,
- the weight of the cylindrical wall, that is, 21,8 t,
- the weight of the sealant (2 cm), that is, 4,25 t,
- the weight of the exterior cover (1 cm), that is, 2,5 t.

That is, the total load of P = 105.5 t.

This vertical force was then broken down into a compression force transmitted by the connecting cone on the tower and a traction force supported by the lower belt of the tank;

– the resulting thrust of the water in the lower element of the tank, that is, 26.2 t.

Since the connecting cone was inclined at 38.40°, the horizontal thrust due to the permanent load P was $Q = P/tg38.40° = 132$ t.

The total thrust was therefore 132 + 26.2 = 158.2 t.

Therefore, per ml of belt: $158.2/\Pi d = 4.57$ t/ml.

This thrust generated traction in the belt of $4.5 \times d/2 = 24.8$ t.

The steel section was then 24,800/12 = 2,066 mm².

The frames could therefore be composed of 4Φ26.

4.7.2.3.3. Checking the inner walls of the peripheral tank

The calculation was carried out in the same way as for the external wall but with a water height of 3.81 m and a radius of 4.60 m. The total section of the hoops then became $A_s = 2,800$ mm².

4.7.2.3.4. Checking the connecting cone

The loads applied to the cone were:

– to the self-weight of the concrete, that is, $0.13 \times 2,500 = 325$ kg/m²;

– to the weight of the sealant, that is, $0.02 \times 2,000 = 40$ kg/m²;

– to the weight of the exterior cover, that is, $0.01 \times 2,000 = 20$ kg/m².

That is, a total of p = 385 kg/m².

This load was often replaced in calculations by an equivalent water height h such that:

$$1{,}000\ h = p \cos 38.40°, \text{ with } h = 0.30 \text{ m}$$

The total water height was then H = 3.64 + 0.30 = 3.94 m.

The calculation of the pressure on the walls of the cone was therefore: P = 1,000H/sin 38.40°.

Hence, a traction T in the concrete wall of:

$$T = 1{,}000\ H \times L \times Dm/(2 \times \sin 38.40°) = 62.72 \text{ t}$$

with:

– D_m: mean diameter of the cone = 9.37 m;

– L: length of the generator of the truncated cone = 2.11 m.

This gave the steel cross-section A_s necessary to balance this tensile force:

$$A_s = 67{,}720/12 = 5{,}230 \text{ mm}^2$$

It was also appropriate to check the compressive stress of the concrete where the cone was connected to the tower.

The total load applied considering the weight of the various structural elements (covering dome, interior wall, cone, water, etc.) was 20.6 t/ml.

The compressive stress was: 20,600/(100 × 13) = 16 bar ≤ 44.8 bar.

4.7.2.3.5. Checking the bottom dome

The calculation was identical to that done for the covering dome. The normal compression at the base of the structure was 17.81 t, that is, a compressive stress of the concrete of 17,810/(100 × 13) = 13.7 bar ≤ 88.8 bar.

The maximum traction in the lower belt of the bottom dome was obtained by considering the case where the outer tank was empty and the inner tank was full.

The tensile force was then 61.6 t/ml, or a steel cross-section of 61,600/12 = 5,133 mm².

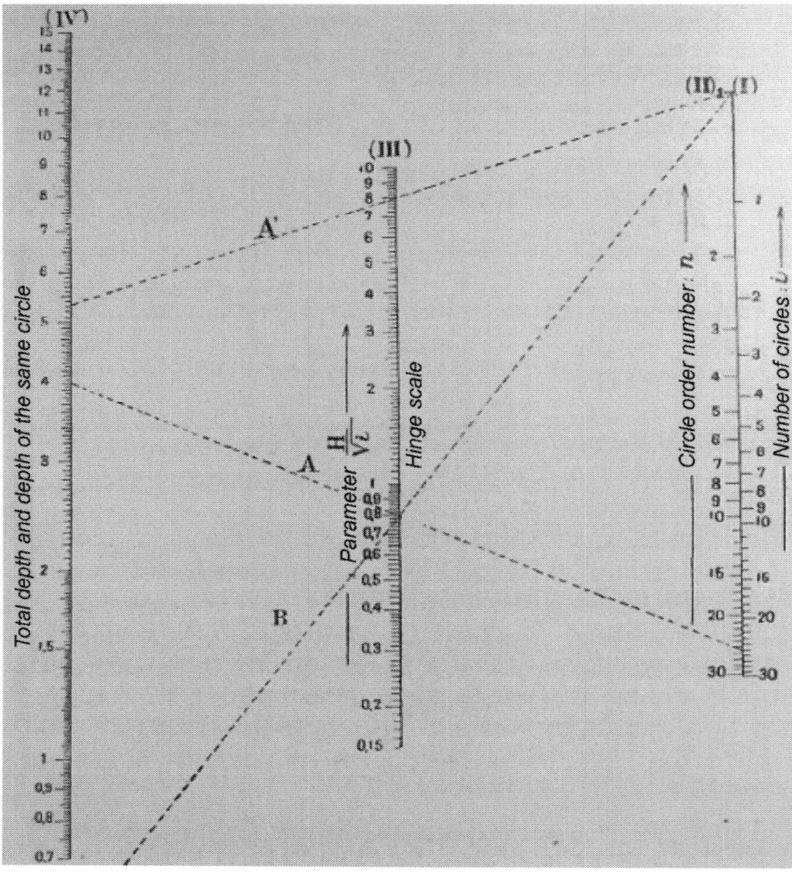

Figure 4.14. *Monogram for the direct calculation of e_n*

NOTE.– In an article published in 1927 in the journal *Le Génie civil*, M. Czyz indicated how to distribute the hoops over the entire height of the reinforced concrete tank so that their working rate was constant. The method used assumed that the hoops, the number of which was equal to i, were all made of round iron bars of identical cross-sections ω. It was assumed that these hoops supported all the tensile forces of the wall. The method then involved dividing the height of wall AB into points i whose distances from the upper limit A were proportional to √i. The tank wall was thus divided into i rings of height (for ring n):

$$\Delta sn = H/\sqrt{i}\,(\sqrt{n} - \sqrt{(n-1)})$$

The center of gravity Gn of the cross-section below the free surface of the liquid was given by the following formula:

$$in = H/\sqrt{i}\ (\sqrt{n} - (\sqrt{n} - \sqrt{(n-1)}))/3 \times (\sqrt{n} - 2\sqrt{(n-1)})/(\sqrt{n} + \sqrt{(n+1)})$$

4.7.2.4. *Santa Eulalia water tank*

In 1928, in the plain of Santa Eulalia in Spain, a specially designed reinforced concrete water tank was built. Since the soil was not of sufficient quality to support a structure on general foundations, the following solution was adopted:

The tank is an open cone shaped bowl whose walls are formed by the assembly of 15 identical, slightly domed sections. The joint between two sections is reinforced by a vertical rib resting on a foundation slab. The slabs together form a 15-pointed star whose center is formed by a circular plate.

Figure 4.15. *Santa Eulalia water tank*

The 15 sectors were built one after the other with the same formwork. For the soles and sectors, the cement dosage was 300 kg/m^3. The ribs were made of unreinforced concrete, and the inside of the tank was coated 4 cm thick in the upper element and 2 cm thick in the lower element.

4.7.3. *The "Pauchot reinforced stone"*

This reinforced stone was in fact reinforced mortar. It consisted of crushed stone, the grains of which had to pass through a sieve of 6 or 8 mm maximum, mixed with a special cement dosed at 250–300 kg/m^3, which did not color the stone. This mortar was iron-railed. The color and appearance of the reinforced stone reproduced those of natural stone. Compressive strength tests of the mortar were carried out by the French National Conservatory of Arts and Crafts on reinforced and unreinforced test specimens.

For a cement content of 250 kg/m^3, the compressive strength of the sample was 118 bar if it was unreinforced and 135 bar if it was reinforced. For a cement content of 350 kg/m^3, these values were 138 bar and 176 bar, respectively.

This was produced identically to that of reinforced concrete with the same formwork and the same shelling. This resulted in a monolithic structure comparable to reinforced concrete but with the appearance of stone, while allowing a varying thickness between 16 and 25 cm.

Figure 4.16. *Example of a building on rue du Mont-Cenis in Paris*

This finish was obtained by the same processes as those used for natural stone. The composition hardened in the air, so to facilitate resurfacing, it began from the bottom when the work had reached at least 1 month of age, which resulted in a hardness of 8. The final hardness varied between 3 and 6.

The advantage of this process was, in addition to the reduced thickness of the structures, providing a monolithic structure, which, in the event of poor terrain, made it possible to reduce the load by 50%.

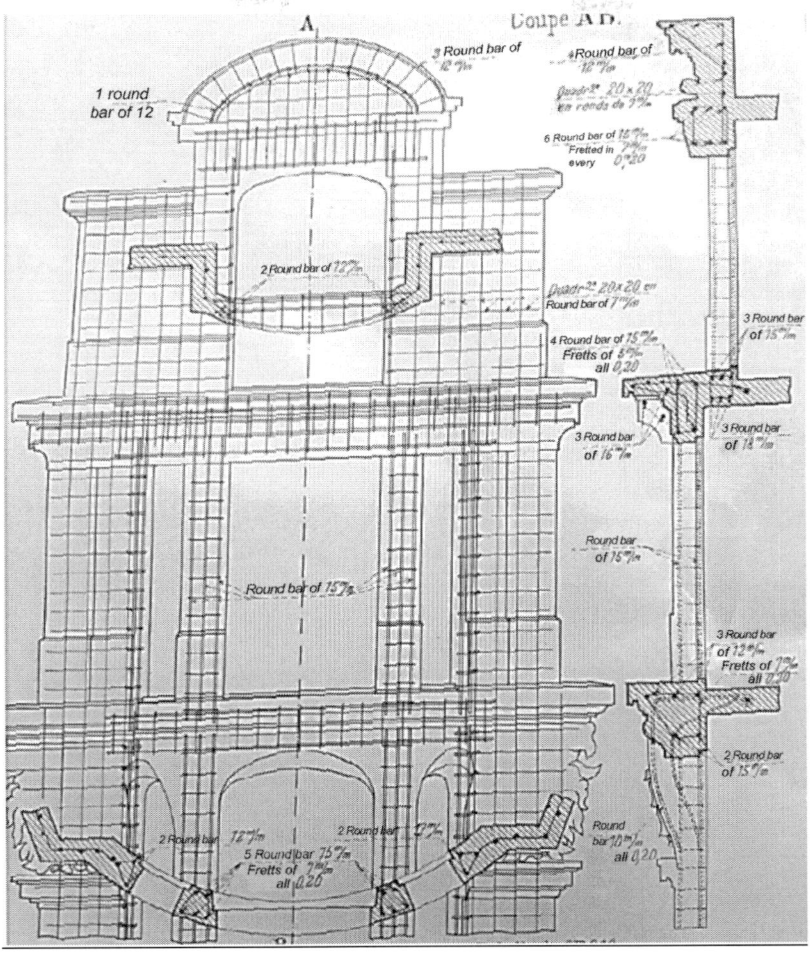

Figure 4.17. *Reinforced stone facade (source: architecture course, E. Arnaud)*

4.7.4. *Hangar for two airships in Luçon (Vendée, France)*

This hangar was designed by Mr. Lossier, engineer, and was to be built by the company Fourré et Rhodes. Adopted by the French General Council of Bridges and Roads from a technical point of view, it was validated by the French Naval Ministry, and the work began when the 1914–1918 war armistice delayed the execution of their plans.

Figure 4.18. *Structural principle of an airship hangar (source: architecture course, E. Arnaud)*

The hangar was in the form of a vault with a free interior height of 53 m, a width of 109 m and a total length of 220 m. To minimize the formwork and scaffolding costs that would have been required for casting in place, Mr. Lossier designed a reinforced concrete frame made from precast voussoirs assembled by a mobile arch. The framework was designed from arches 10 m apart from each other from axis to axis and joined by purlins identical to those of the Montebourg hangar. Each arch was composed of 14 prefabricated 10-m long voussoirs and two abutments (Vierendeel type) cast on site. The height of the voussoirs varied from 2.40 m at the keystone to 3.50 m at the base. Transverse stability was ensured by two lattice girders twinned by hollow core diaphragms. The ends of the members and diagonals were fitted with threaded end bars, which were used to ensure the connection of the various sections. The joints were secured with molten cement. Each truss was therefore considered an arch recessed at its origin without articulation. The loads considered were mainly linked to the wind.

The engineer considered the following:

– pressure of 150 kg/m^2 of surface normally struck at an angle of 10° with the horizontal;

– pressure of 150 kg/m^2 acting on one side of the hangar while the other was subjected to a depression of 50 kg/m^2;

– pressure of 100 kg/m^2 acting on a slope and depression of 100 kg/m^2 on the other slope;

– internal pressure of 85 kg/m^2 in lifting.

Despite its dimensions, the lightness was such that the permanent loads were comparable to those of a 0.23-m-thick solid arch hangar.

5

The Ministerial Guidelines of July 19, 1934 on the Use of Reinforced Concrete

5.1. Overview

This regulation applied to structures under the responsibility of the French Ministry of Public Works.

It aimed to update the guidelines from October 20, 1906, considering new knowledge on construction sites and in laboratories.

It was preceded in 1928 by a draft regulation from the Trade Union of Reinforced Cement Manufacturers. This draft regulation then served as the basis for establishing the new regulations.

The 1906 guidelines were updated by a special commission and then validated by the French General Council of Bridges and Roads. At the same time, the General Council also proposed to modify the general specifications of reinforced concrete on October 29, 1913. Finally, specific cost estimation articles were updated.

The new regulations included three appendices:

– Appendix I: new instructions on the use of reinforced concrete in structures under the Ministry of Public Works (replacing those of October 20, 1906, amended by the May 19, 1927 guidelines).

– Appendix II: all articles of the general specifications related to reinforced concrete.

– Appendix III: amending texts made to articles relating to reinforced concrete in specific and processing cost estimation:

- sand for mortar and concrete;
- special provisions for grinding sand;
- stones;
- lime and cements;
- reinforced concrete structures.

Appendix I comprises 17 articles as follows:

– Article 1: armatures;

– Article 2: concrete;

– Articles 3–5: project preparation;

– Article 6: shrinkage;

– Articles 7–10: structural calculations;

– Articles 11–15: constructive provisions;

– Article 16: execution of works;

– Article 17: tests.

The following symbols were adopted:

– R_b: compressive stress of the concrete;

– R_a: compressive stress of steels;

– N: compressive force;

– Ω: cross-sectional area;

– Ω_a: cross-sectional area of the steel;

– Ω_b: cross-sectional area of the concrete;

– E_a: modulus of elasticity of the metal;

– E_b: modulus of elasticity of the concrete;

– $m = E_a/E_b$: density;

– M: bending moment;

– I_t: moment of inertia of the total cross-section.

5.2. Calculation bases: requirements

5.2.1. *Basic requirements*

The following principles were considered:

– the "admissible stress" calculation was carried out;

– loads were considered with their characteristic values.

The calculations were performed without a partial coefficient. The greatest live loads that the works will have to support should be considered (for buildings, reservoirs, pipes, etc.).

Bridges were subject to live loads of metal bridges according to the regulation of August 29, 1891 (replaced by that of May 10, 1927).

5.2.2. *Project usage duration*

The project usage periods were not mentioned in the regulation.

5.2.3. *Thermal effects*

There are no guidelines on the values of possible thermal effects and how to take them into account.

Article 5 specifies, however, that the temperature difference to be considered was ±20°C.

5.2.4. *Settlements and differential movements*

There are no specific data on this subject in the regulation.

5.2.5. *Material properties*

The regulation does not include guidelines for considering concrete creep.

5.3. Materials

5.3.1. *Concrete*

5.3.1.1. *Resistance*

Article 2 of the regulation limited the value of the mechanical resistance of concrete to compression (the so-called "fatigue" limit) to:

$$28/100 \times \text{compressive strength at break at 90 days}$$

According to the text, the concrete test specimens were cubes of size 20 × 20 × 20.

The text also stated that the tensile, shear and adhesion strength were limited to:

$$20/100 \times \text{tensile strength at 90 days}$$

NOTE.– The text recommended the use of 20/25 cement for large structures.

It was also recommended to choose concrete with the following compositions:

– 400 L of sand with a maximum dimension of 5 mm;

– 800 L of gravel with a maximum dimension of 25 mm;

– 300, 350 or 400 kg of 20/25 class cement.

Plasticity corresponded to subsidence between 2 and 10 cm. The results are summarized in Table 5.1.

Cement content (kg/m^3)	Compressive strength at 90 days (MPa)	Tensile strength at 90 days (MPa)	Compression fatigue rate (MPa)	Fatigue rate in traction, shear, adhesion (MPa)
300	21.5	3	6	0.6
350	23.5	3.25	6.5	0.65
400	25	3.5	7	0.7

Table 5.1. *Concrete strength values*

5.3.1.2. *Elastic strain*

The modulus of elasticity of the concrete was calculated from the characteristic diagram of the concrete (stress–strain diagram). The elastic modulus of the steel was equal to E_a = 220,000 MPa. The modulus of concrete is less well known because its variability depends on its composition and age. The extreme limits ranged from 15,000 to 40,000 MPa.

The steel/concrete equivalence coefficient, referred to as the density m = E_a/E_b, varies from 8 to 15 and is commonly taken as 10.

5.3.1.3. *Shrinkage*

Article 6 of the regulation defined for the first time the effects of concrete shrinkage (the 1906 regulations did not mention it). Shrinkage was assimilated to inelastic shortening whose value was fixed at 2×10^{-4} m/m. However, to partially reduce the effects of the latter, the following provisions were considered:

– use of low shrinkage concrete (without specifying the nature of this concrete);

– phased works with differential concreting of assembling joints (at supports, for example).

Shrinkage was considered together with the permanent, variable loads, wind and temperature. For the load combinations, the value of the shrinkage was then fixed at 3×10^{-4} m/m.

NOTE.– At the time, the phenomenon of concrete shrinkage was relatively poorly understood.

It was often considered that this phenomenon could continue for several years after hardening with periods of stop and even regression of shrinkage depending on the atmospheric conditions and the loads applied. It was also accepted that it was dependent on:

– the composition of the concrete (the granulometry of the "inert" materials, the nature of the cement matrix and the quantity of mixing water);

– the amount of concrete poured (length, height, etc.).

5.3.2. *Reinforced concrete steel*

5.3.2.1. *Properties*

Article 1 of the regulations referred to the guidelines of May 10, 1927 on metal bridges. Nothing prohibited the use of other types of steels if cracking in tensioned concrete could be controlled. It was also possible to use mild steel with a lower fatigue limit (but not more than 10%).

5.3.2.2. *Resistance*

The admissible "fatigue" limits considered are summarized in Table 5.2.

Load	Fatigue limit of steel in tension or compression (MPa)	Fatigue limit of steel in shear or sliding (MPa)
Permanent loads + live loads (increased dynamic effects) + temperature	130	104
Permanent loads + live loads (increased dynamic effects) + temperature + wind (150 kg)	140	112
Permanent loads + temperature + wind (250 kg)	140	112

Table 5.2. *Load combinations*

The ministerial guideline of February 7, 1933 permitted the use of medium-hard steels (called "high-strength steels") in round bars (Ac-54) with a strength at break of between 54 and 64 kg/m^2 (540–640 MPa) and an elastic limit greater than 36 kg/mm^2 (360 MPa). The working rate of this type of reinforcement was set at 18 kg/mm (180 MPa) for an elongation of 0.65 mm/m.

5.3.2.3. *Steel/concrete adhesion*

These provisions were developed in the 1930 regulation, and the resistance to sliding bars in concrete was given by the following formula:

$$0.2/(1 + d_1/e_1)\ (1 + d_1/e_2)$$

In this formula, d_1 is the diameter of the bar, and e_1 and e_2 are the minimum distances from the axis of the bar to the free surface of the concrete in the two perpendicular directions.

Overall, for concrete with a compressive strength of 200 bar (20 MPa), the cohesive strength was 40 hectopiezes (4 MPa) for a bar located at the center of the concrete and 10 hectopiezes (1 MPa) for a bar located at a distance from the surface equal to its diameter.

For more information, the EC 2 values are given in Table 5.3.

fck (MPa)	12	16	20	25	30	35	40	45	50	60
$l_{b,rqd}/\Phi$ good adhesion	64	54	47	40	36	32	29	27	25	24
$l_{b,rqd}/\Phi$ poor adhesion	91	78	68	57	52	45	42	39	36	34

Table 5.3. *Values of the overlap lengths as a function of the mechanical resistance of the concrete according to EC 2*

In this table, $l_{b,rqd}$ represents the overlap length and Φ represents the diameter of the reinforcements.

5.4. Durability and cover of reinforcements

5.4.1. *Overview*

The sustainability criteria are not mentioned in the regulations, nor are the conditions of exposure to physicochemical attacks.

5.4.2. *Durability*

Lifespan is not mentioned.

5.4.3. *Cover*

Article 12 from the regulation gives cover values of *20 mm* for current structures and *35 mm* for structures at sea. The commentary specified, however, that the distances between reinforcements had to be greater than C_g + 5 mm, where C_g was the diameter of the largest aggregate.

The mechanical cover was set at a value greater than or equal to the diameter of the reinforcement.

5.5. Structural analysis

5.5.1. *Overview*

The purpose of structural analysis is to determine the distribution of stresses or pressure and strain. The behavior model used is that of *linear elastic behavior*. *Article 7* states:

> Deformation and strength calculations will be derived from classical principles of material strength or equivalent scientific principles.

5.5.2. *Load combinations*

By default, Article 9 of the 1906 regulation was extended and requested that the resistance calculations consider:

– the greatest external forces;

– wind;

– snow;

– thermal strain (for non-freely expandable structures);

– shrinkage (for non-freely expandable structures).

These loads were combined without a partial coefficient.

Article 1 mentioned three combinations in the table of admissible constraints of steels:

– combination 1: permanent load + live load (increased by possible dynamic effects) + temperature;

– combination 2: permanent load + live load (increased by possible dynamic effects) + temperature + wind at 150 kg;

– combination 3: permanent load + temperature + wind at 250 kg.

5.5.3. *Second-order effects*

Second-order effects that can significantly affect the *overall stability* of the structure were not considered in the regulation.

5.5.4. *Structural modeling*

There were no suitable digital means for modeling structures.

5.6. Structural calculations

5.6.1. *Simple compression*

The simple compression calculation was carried out using the fictitious homogeneous cross-section called Ω.

The value of Ω was given by:

$$\Omega = \Omega_b + m\,\Omega_a$$

with:

– $\boldsymbol{\Omega_b}$: concrete cross-sectional area;

– $\boldsymbol{\Omega_a}$: total cross-sectional area of longitudinal reinforcements;

– **m**: steel/concrete equivalence coefficient.

The calculation of the compressive stress on the concrete was as follows:

$$R_b = N/\Omega$$

That of the constraint on steel is as follows:

$$R_a = m\,N/\Omega$$

The coefficient m was generally equal to 10, but it was possible to take a different value depending on the quality of the concrete.

5.6.2. *Compression with bending*

Article 9C of the regulations specified that in the case where the material resistance rules applied to the homogeneous cross-section, the absence of traction in the section was considered, and the actual distribution of normal stresses in this section respected the dimensioning principles.

Otherwise, a reduced cross-section was considered, with no tension in the concrete. This was calculated in the same way as that of the 1906 regulations, considering:

– the area Ω of the fictitious cross-section (homogeneous section);

– the moment of inertia relative to the transverse axis of bending $I = I_b + m\ I_a$.

The constraints were then given by the classical formulas for the resistance of materials, namely:

– *concrete compression on the extreme fibers*:

$R_b = N/\Omega + (M\ v_b)/I$, with v_b distance from the extreme fiber to the axis of inertia

The formula was modified to consider the slenderness through the radius of gyration r ($I = r^2\ \Omega$) according to:

$$R_b = N/\Omega\ (1 +/- (v_0 v_b/r^2))$$

where $\mathbf{v_0}$: eccentricity of N relative to the geometric axis;

– *compression of steels on extreme fibers*:

$R_a = m\ (N/\Omega + (M\ v_a)/I)$, where v_a is the distance from the extreme steel to the axis of inertia

The effect of the shear force was calculated on the previous reduced surface with the classic material resistance rule formulas.

5.6.3. *Bending*

The case of bending is described in section 5.6.2.

5.6.4. *Buckling*

Article 10 of the regulations validated the fact that elements whose slenderness was less than 20 and whose concrete compressive stress remained below the limit value defined in Article 2 (Table 5.1) did not have to be checked for buckling.

Otherwise, Rankine's formula was customarily used:

$$N/\Omega\ (1 + kl^2/10{,}000r^2) \leq R_b$$

with:

– N: total compressive force;

– l: length of the element;

– r: radius of gyration;

– k: coefficient depending on the conditions at the ends according to Table 5.4.

Conditions at the ends	Value of k	Observations
An embedded support **The other free**	4	–
An embedded support **The other articulated**	0.5	In case of imperfect embedding $0.5 \leq k \leq 1$
Two articulated supports	1	–
Two embedded supports	0.25	In the event of imperfect embedding of one of the supports: $0.5 \leq k \leq 1$ In the event of imperfect embedding of both supports: $0.25 \leq k \leq 1$

Table 5.4. *Values of k according to support conditions*

In the case of an element undergoing composite bending (bending M + compression N), the Rankine formula was replaced by:

$$N/\Omega\ (1 + kl^2/10{,}000r^2) + Mv/I \leq R_b$$

or:

$$N/\Omega\ (1 + kl^2/10{,}000r^2 + v_0 v/r^2) \leq R_b$$

5.6.5. *Example of simple compression calculation*

The data are as follows:

– rectangular post 400 × 600;

– permanent loads G = 1.5 MN;

– variable loads including temperature Q = 0.5 MN;

– concrete quality: mechanical resistance to crushing at 90 days of 25 MPa, cement content of 400 kg/m^3;

– buckling length l = 4 m;

– Fe 24-grade steel, 8 cm^2 steel section.

The total compression acting on the post is as follows:

$$N = G + Q = 2 \text{ MN}$$

The compressive stress supported by the concrete is as follows:

$$R_b = N/\Omega = 2/(0.4 \times 0.6) = 8.33 \text{ MPa}$$

The compressive stress supported by the steel is as follows:

$$R_a = m\ N/\Omega = 10 \times 2/(0.4 \times 0.6) = 10 \times 8.33 = 83.3 \text{ MPa}$$

These values are compared with the values of the above-mentioned admissible constraints, that is,

– 7 MPa for concrete;

– 130 MPa for steel.

The concrete cross-section is too small.

5.7. Constructive provisions

The main provisions relating to the execution of the works were included in Chapter III of Appendix III of the regulations under the title: “Execution of the works”. As these were not very detailed, the 1930 regulations established by the Trade Union Chamber were generally used.

5.7.1. *Concrete pouring*

The layers were “compacted” or “pounded”, the effect being optimal when the water flowed back to the surface. The use of “shaking tables” or vibrators was also considered.

5.7.2. *Distance between reinforcements and between reinforcements and the formwork*

The distance between the steel as well as between the steel and the formwork had to allow for:

– "ramming" of the concrete;

– its tightening against the frames;

– a minimum cover of 20–35 mm to meet the durability criterion.

5.7.3. *Construction joints treatment*

The construction joints treatment is as follows:

– "raw" cleaning of the old concrete;

– wetting of the recovery surface;

– pouring fresh concrete.

NOTE.– The use of cement slurry was not recommended.

5.7.4. *Note No. 1: Appendix II*

The rolled steel had to be subjected to the following tests:

– tensile tests applicable to all steel;

– cold bending tests applicable to all steel;

– cold punching tests applicable to sheets, plates and profiled sections;

– hot bending tests applicable to sections;

– brittleness tests for steel bars.

The results of the tensile tests were considered satisfactory if they met the requirements presented in Table 5.5.

The number of test pieces to be produced was set at:

– 2 by 3 t up to 20 t;

– 2 by 4 t over 20 t.

– Composition and manufacture of ordinary concrete: ordinary non-reinforced concrete was composed of a 3:2 ratio of rubble to mortar. The minimum cement content of the reinforced concrete was set at 250 kg/m^3. The cement had to meet the general conditions set by the March 13, 1928 regulation revised on December 18, 1933.

– Implementation of reinforced concrete steel: the reinforcements had to be clean without traces of non-adherent rust, paint or grease. The implementation tolerances compared to the plans were limited to the minimum values:

- $\Phi/2$ (Φ is the diameter of the bar or the thickness of the plate);
- 6 mm.

Name	Elastic limit (kg/mm^2)	Breaking limit (kg/mm^2)	Elongation (%)
Sheets, plates, round, square or profiled bars	24	42	25
Checkered sheets	20	40	20

Table 5.5. *Limit values of the resistance of reinforced concrete steel*

The overlap of the bars was, in the zones under tension, ensured by a length calculated according to the steel/concrete adhesion, and the steel was finished with hooks.

– Concrete strength testing and control: Compressive strength tests were carried out on 20 × 20 × 20 cubic specimens at 90 days. Tests at 7 and 28 days were also planned.

5.7.5. *Note No. 2: Regulation on reinforced concrete constructions (emanating from the Trade Union Chamber in January 1930)*

The report of the Trade Union Chamber upon the 1906 Instruction was composed of four titles.

Title I: this title indicates the preliminary data necessary for the development of any construction. In particular, the live loads are considered in the calculations. These values are summarized in Table 5.6.

NOTE.– It was also requested that a concentrated load equal to 2S2 be considered on 1 dm^2 of cover.

The loads were taken at the maximum value and not at the average value.

The combination of loads to be considered for the calculation was as follows:

Permanent load + $S_1 + 4/3\ S_2 + S_3$

S_1 constituted fixed loads (not highly variable over time) but could occupy any position. S_2 were loads variable over time. S_3 was dynamic loads determined by inertial forces.

Title II: this title dealt with execution drawings and plans.

Title III: this concerned the materials used in reinforced concrete. All types of steel were permitted, including structural steel profiles that could be assembled by bolting or riveting before being implemented in the concrete. With respect to binders, the use of special cement such as aluminous cement was permitted under certain conditions. For the aggregates, the maximum diameter of the aggregate had to be less than half the distance between reinforcements in the same layer.

Premise	Fixed live load S1 (kg/m²)	Variable live load S2 (kg/m²)	Dynamic live load S3 (kg/m²)
Private residence	100	100	0
Offices where the public is only admitted in isolation	100	200	0
Small shops	100	200	50
Performance halls, dance halls, etc.	100	300	100
Workshops	100	400	200
Heavy-duty workshops	200	800	400
Warehouses	100–300	400–1,900	200–400

Table 5.6. *Values of variable charges depending on the destination of the premises*

The mechanical resistance of the concrete recommended by the regulation was:

– 160 bar (16 MPa) for small structures;

– 200 bar (20 MPa) for more important structures;

– 250 bar (25 MPa) for very important structures.

Title IV: this title mentioned strength calculations. Article 22 specified that "cracking of the areas of the concrete under tension should be considered as a normal phenomenon in the operation of the concrete–steel system" (provision reused in Chapter 7 of EN 1992.1.1).

The density of reinforced concrete in the calculation of permanent loads was estimated at 2.5 t/m^3.

The usual calculation values are summarized in Table 5.7.

Calculation parameters	Values	Comments
Modulus of elasticity of concrete (1)	*With CPA preserved in water*: $E_i = E_v$ = 1,000 myriapiezes (10,000 MPa) *With CPA under normal conditions*: E_i = 3,000 myriapiezes (30,000 MPa) E_v = 1,000 myriapiezes (10,000 MPa) *With aluminous cement*: E_i = 4,500 myriapiezes (45,000 MPa) E_v = 1,500 myriapiezes (15,000 MPa)	In certain cases, manufacturers were allowed to make a study of actual strain based on experience. The average value of 2,000 myriapiezes (20,000 MPa) was allowed for calculations except for buckling.
Poisson coefficient	0.15	
Shrinkage	Average value of the shrinkage coefficient was 2×10^{-4} Except below the Bayonne–Grenoble line where its value is 3×10^{-4}	For concrete in contact with the ground or a saturated atmosphere, typically shrinkage is not considered. For concrete in contact with a sheltered and heated atmosphere, the coefficient was then 6×10^{-4} with E = 1,000 myriapiezes (10,000 MPa).
Variations in temperature	±12°C	This value increases to ±20°C in leaflet 74 of 1998.

(1) E_i is the instantaneous modulus of the concrete and E_v is the deferred module.

Table 5.7. *Usual values of the parameters for sizing*

5.8. Comments

5.8.1. *General comments*

To increase the mechanical resistance of steel, some manufacturers had the idea of taking two mild steel bars and twisting them, as shown in Figure 5.1.

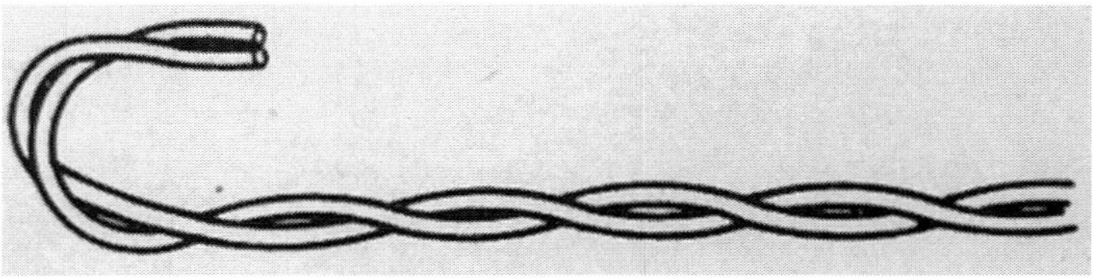

Figure 5.1. *Isteg steel. For a color version of this figure, see www.iste.co.uk/lauzin/concrete.zip*

This steel had the mechanical characteristics summarized in Table 5.8.

Characteristics	Values
Breaking strength	40 kg/mm^2
Elastic limit	36 kg/mm^2
Elongation at break	10%
Allowable stress	18 kg/mm^2

Table 5.8. *Mechanical characteristics of Isteg steel*

The steel/concrete adhesion increased by 25% compared to a conventional mild steel. Compared with mild steel, which is more rigid, Isteg steel is safe for cracking under the same tensile stress.

In 1935, in his treatise on masonry, Georges Debès (chief engineer of bridges and roads) mentioned the possibility of "putting the reinforcements under pre-tension, which transfers an initial compressive stress to the entire mass of concrete". This process was envisioned by constructors such as Considère, Bach and Morsh. The idea of putting the steels under tension of the same order as the traction undergone by the element then came up against the problem of considering the effects of shrinkage, elastic shortening of the concrete and creep (called "progressive deformation under constant load"). This last phenomenon was highlighted by Freyssinet and Faber in 1927.

Shrinkage cracking: It was typical in certain hot regions (southeast France in particular) to mix concrete with excess water to counteract the effects of evaporation. This excess water generated shrinkage cracking (even during setting). At the time, one of the known effects of using a calcium chloride-based water repellent was the increase in shrinkage phenomena. The solutions used to prevent shrinkage were as follows:

– do not force the cement dosages;

– make plastic concrete but without too much water;

– tighten well when pouring;

– do not subject the element to too rapid drying (protect from sunlight, keep the concrete humid for two weeks), and *fill the tanks with water as soon as the concrete hardens*.

Impermeability: Concrete is typically considered impermeable (and not watertight) for thicknesses of 3–8 cm with a cement content between 400 and 600 kg. However, for hydraulic structures (reservoirs, dams, etc.), a waterproofing layer was added either as a water-repellent mortar (generally bitumen) or non-water-repellent, which was covered with a layer of bitumen or tar paint, or in reinforced concrete 10 cm thick protected by a bituminous cover. It is noted in the treatise of Georges Debès that "the incorporation of so-called water-repellent materials into concrete to increase its impermeability must be formally prohibited in reinforced concrete".

Actions of external agents: In atmospheres aggressive toward concrete, the following is recognized:

– sulfated water (aggressive for Portland cements);

– acidic waters;

– acid fumes.

It was then decided to protect the concrete and the reinforcements in the following way:

– extra thickness of the concrete to protect the steel;

– cover concrete with molten cement;

– painting concrete with special paints;

– implement a protective device (e.g. under the slabs of bridges exposed to train smoke, a sacrificial slab in reinforced ceramic was used);

– for tanks likely to contain aggressive liquids (acidic, syrupy and oleaginous), use special paints or covers (concrete wine tanks, for example, could be covered with glass slabs);

– spraying cement (cement gun) was considered a solution for repairing damaged structures, and the cement content was between 500 and 600 kg of cement per m^3;

– concrete for maritime use: builders usually consider sea water to be unable to attack Portland cement-based concretes (despite the presence of sulfates in this water), and compact concrete was sufficient. Only a thicker *35 mm* cover of the steel was used.

5.8.2. *Comparison between the 1906 and 1934 regulations*

Articles	1906 regulations	1934 regulations	Comments
Steel quality	Not specified	*Normal quality*: mild steel whose characteristics are: breaking strength 42 kg/mm² elastic limit 24 kg/mm² elongation at break 25% *Low quality*: tolerance of –10% on these characteristics *Hard quality*: freedom to use them provided that the concrete is checked for cracking	
Allowable tensile stress of steel	*Common cases*: 50% elastic limit *Cases of fatigue or shock*: 40% elastic limit	Under the load combination (1): CP + Q + T the value is 13 kg/mm² CP + Q + T + Wn the value is 14 kg/mm² CP + T + We the value is 14 kg/mm²	*For lower quality steel,* the stress was reduced in the ratio of the breaking strengths *For hard steel*, the stress was limited to 50% of the elastic limit without exceeding 30% of the breaking strength

Concrete quality	Not specified It was mentioned the use of concrete dosed at 300, 350 and 400 kg of Portland cement (compressive strength of 16, 18 and 20 MPa, respectively) No tensile strength value	Not specified Same dosages as in 1906 For large structures, the use of Portland cement 20/25 was required with concrete compressive strengths at 90 days of: 21.5, 23.5 and 25 MPa. And at a traction of: 3, 3.25 and 3.5 MPa	
Allowable compressive stress of concrete	*For simply reinforced elements:* 28% resistance at 90 days *For fretted elements*: 60% resistance at 90 days	As per 1906	
Steel/concrete adhesion stress	10% compressive stress	20% of the tensile breaking stress at 90 days	
Structural calculation principles	Calculation of admissible stress Maximum load values	As per 1906	
Shrinkage	Not specified	The shrinkage coefficient was set at 2×10^{-4} m/m The combined effect of temperature and shrinkage was considered with a total shrinkage of 3×10^{-4} m/m	
Creep	Not specified	Not specified	

(1) CP: permanent charges; Q: live load increased by possible dynamic effects; T: effect of temperature; In: wind stress at 150 kg/m²; We: wind stress at 250 kg/m².

Table 5.9. *Design rules*

Articles	1906 regulations	1934 regulations	Comments
Steel cover	Minimum 15–20 mm	Minimum diameter of steel for mechanical cover Minimum of 20 mm for ordinary structures and 35 mm for maritime structures for durability	The condition Cg + 5 mm was also applicable
Distance between steels	Distance to allow the implementation and tightening of concrete between the reinforcements	Diameter of the largest gravel + 5 mm (Cg + 5 mm)	
Bending the bars	Not specified	Diameter of hooks = 5d (d: bar diameter)	It was further requested that the hooks be held by frets to counteract lateral reactions
Arrangements of reinforcements for slabs and joists	For a working direction, two perpendicular layers: – reinforcements calculated in the load-bearing direction – 50% of the load-bearing steels in the distribution direction	As per 1906 plus minimum values of the "mesh"	
Implementation methodology	– Rigid formwork – Slow-setting cement – Concrete compacted in 5 cm sections – Treatment of construction joints by cleaning and wetting – Keep the concrete moist for 15 days after pouring – Remove formwork when the required strength is reached, without shocks	As per 1906, place bars in the construction joints to ensure the transmission of shear forces	

Table 5.10. *Constructive provisions*

5.8.3. *Practical data*

The steel most commonly used in reinforcements consisted of round bars of 8–30 mm; profiled iron was sometimes used in large foundation structures or posts; transverse reinforcements ("stirrups", frames, etc.) consisted of:

– of small-diameter round iron bars;

– of strips (flat iron) with a width of between 20 and 30 mm and a thickness varying from 1.5 to 3 mm.

Cement characteristics	**Maximum diameter of aggregates (mm)**	**Cement dosage for 16 MPa concrete (kg/m³)**	**Cement dosage for concrete 20 MPa (kg/m³)**	**Cement dosage for concrete 25 MPa (kg/m³)**
Cement 15/20	25	290	335	390
	20	310	355	420
	15	330	385	450
	10	365	420	495
	5	435	500	590
Cement 20/25	25	–	280	335
	20		300	355
	15		320	385
	10		350	420
	5		425	500

Table 5.11. *Common composition of concrete*

The percentage of reinforcement (the ratio of the volume of metal to the volume of concrete) in practice was as follows:

– 0.5%–1% for weakly reinforced structures;

– 1.5%–2.5% for normally reinforced structures such as buildings and industries;

– 3%–4% for heavily reinforced structures;

– 5% for very heavily reinforced structures such as fretting areas, for example.

The density of the reinforced concrete was 2,500 kg/m^3. This value corresponded to a reinforcement density of 1%–2%.

The 1930 regulation of the Trade Union Chamber considered higher admissible constraints than those mentioned in the 1934 regulation (Table 5.12).

Notably, however, the 1930 rules used a combination of factored loads, which was not the case in the 1934 regulations.

Materials	Permissible constraints according to the 1930 regulation	Permissible constraints according to the 1934 regulation
Steel	CP + Q combination: 53% × elastic limit Full combination: 60% × elastic limit	Under CP + Q combination: 54% × elastic limit Under full combination: 58% × elastic limit
Concrete	CP + Q: 32% × breaking limit Under full combination: 60% × breaking limit	Under full combination: 28% × breaking limit

Table 5.12. *Comparative table of admissible stresses of materials between 1930 and 1934*

5.8.4. *Examples of the Boussiron hangars*

Figure 5.2. *View of the facade of a Boussiron hangar (upper left stabilizing support leg). For a color version of this figure, see www.iste.co.uk/lauzin/concrete.zip*

World War I brought opportunities to the field of reinforced concrete in the context of programs for the construction of large arms factories and airship hangars.

For the latter, Henry Lossier built remarkable structures in Montebourg between 1916 and 1919 and then in Le Bourget in 1921. Other noteworthy structures were the twin hangars at Orly built by Eugène Freyssinet and the Limousin Company between 1921 and 1923.

The reinforced concrete roofs of these hangars offered spans of 50 m. Most of the constructions included a vaulted slab stiffened by upper or lower ribs extended to the ground by posts. The thrust of the vault was then balanced by braces or tensioners located at the top of the posts and connected to the ribs by vertical uprights or reticulated systems.

In Marignane, France, Nicolas Esquillan and the company Boussiron began building hangars in 1938, achieving a record first free span of 80.50 m and then a second of 101.50 m (double hangars completed in 1951). This chapter describes the latter.

The structures of these hangars are based on the technology of thin reinforced concrete plates. To reach the 100 m span required in the specifications, Nicolas Esquillan imagined a *thin toric vault that was* 6 cm thick and divided into six waves.

No waterproofing cover was added to the vaults since waterproofing was ensured according to the design at the time (by the compression of the concrete, the compactness of the vibrated concrete and by enriching the extrados of the waves by sprinkling fresh concrete containing 2–3 kg of cement).

Figure 5.3. *Thin stiffened shell cover with reinforced concrete tension rods. For a color version of this figure, see www.iste.co.uk/lauzin/concrete.zip*

Each section is stiffened by transverse slabs and supported by reinforced concrete tension rods. The span of the roof is thus 101.50 m with a spire of 12.10 m.

In addition, the structure has a skewback to strengthen the supports where the posts intercept the tension rods.

Another particularity of the hangar is the glass facades whose panes are held in place by reinforced cement joinery.

Figure 5.4. *Formwork removal*

To build the structure, Nicolas Esquillan jacked the roof at the level of the posts. These prefabricated U-shaped posts were used as assembly columns. A daily height of 1.06 m was selected for future lifting; after each day, a section of the post was completed by concreting according to the vacuum concrete process.

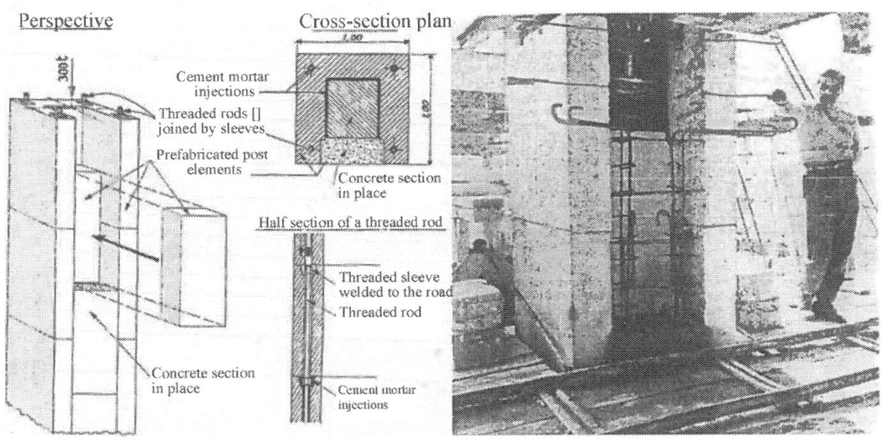

Construction of the posts

Closure wall. Strengthening of the pending steel bars. Reinforcement.

Figure 5.5. *Post serving as a mounting column*

During the temporary lifting and installation phases, it was necessary to create wind bracing elements (in the absence of braces and wind bracing walls). This role was fulfilled by adjustable steel pads sliding on metal plates. It took 38 days to assemble the first cell.

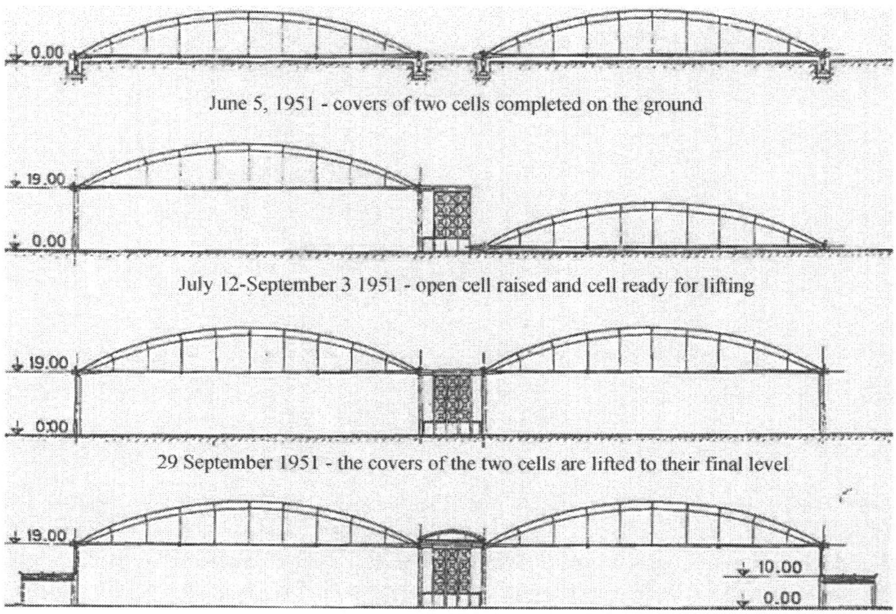

Figure 5.6. *Cell assembly kinematics*

Figure 5.7. *Coverage of both cells completed. French Institute of Architecture, funds by Perret 535 AP 660-15*

5.8.4.1. *Analysis carried out in 2012 by LERM*

In 2012, approximately 60 years after its completion, the structural concrete was assessed. Samples were taken by LERM in five areas: two on the braces and three on the thin roof vaults. The results are summarized in Table 5.13.

Measurement	Measured values	Regulatory values (1934)
Cement content (kg/m^3)	350	350
Porosity	15%	Not specified
Mechanical resistance	Between 22 and 34 MPa	23.5
Aggressive chlorides	0	–
Steel cover	15 mm	15–20 mm (increased to 35 mm by the Trade Union Chamber for maritime structures)
Carbonation depth	40 mm	–

Table 5.13. *Characteristics of the concretes collected; comparison with regulatory values*

5.8.4.2. *Verification of structures*

The brace and post: the detailed plan of the reinforcement of the post and the brace is shown in Figures 5.8–5.10.

Concerning the brace, the reinforcement was composed of (section 2.2):

– 14 × 32 diameter steels in the upper part;

– 14 × 32 diameter steels in the lower part.

Concerning the post, the reinforcement was composed of (section 1.1):

– 10 × 32 diameter steels in the upper part;

– 10 × 32 diameter steels in the lower part.

Capacity of the brace and the post: simple compression was calculated from the fictitious homogeneous section Ω. The value of Ω was given by:

$$\Omega = \Omega_b + m\ \Omega_a = 0.60 \times 0.60 + 8 \times 0.0113 = 0.45\ m^2$$

with:

– $\mathbf{\Omega_b}$: concrete section area;

– Ω_a: total area of longitudinal reinforcement sections;

– **m**: steel/concrete equivalence coefficient.

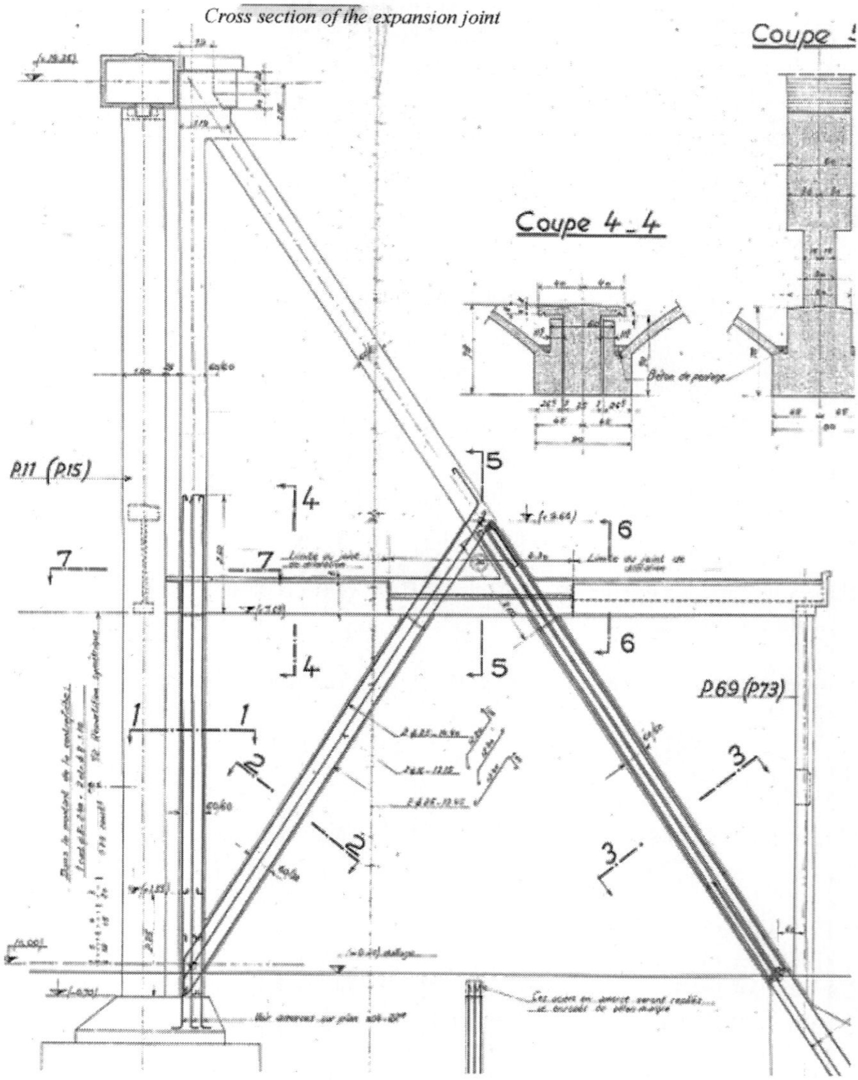

Figure 5.8. *Reinforcement plan for the post and the brace*

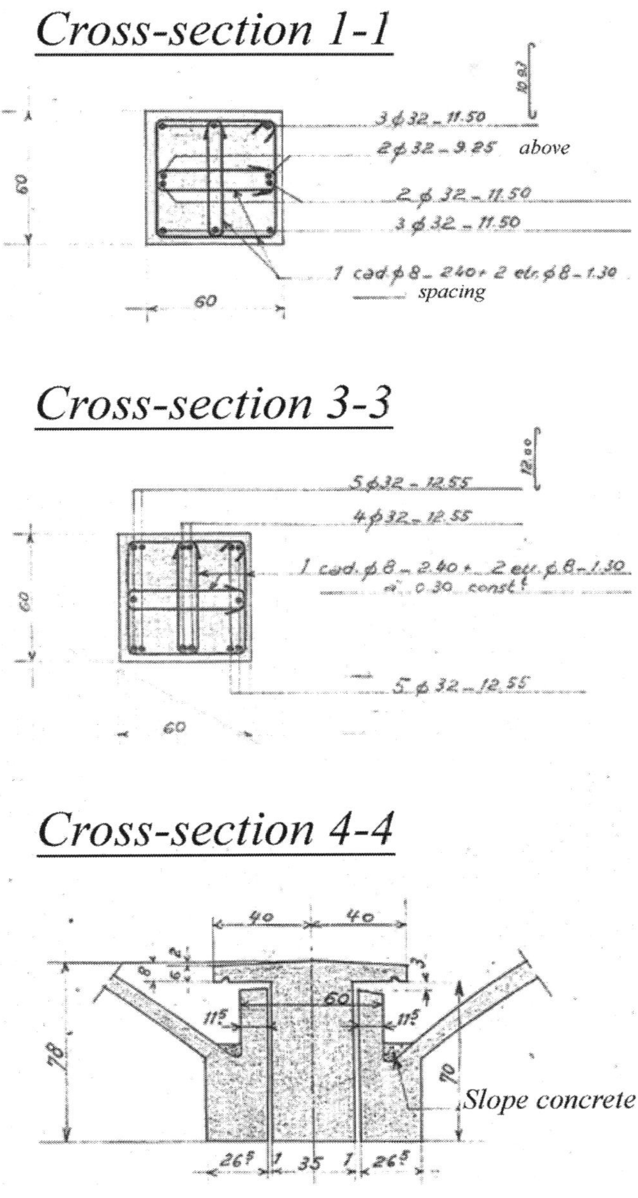

Figure 5.9. *Reinforcement details*

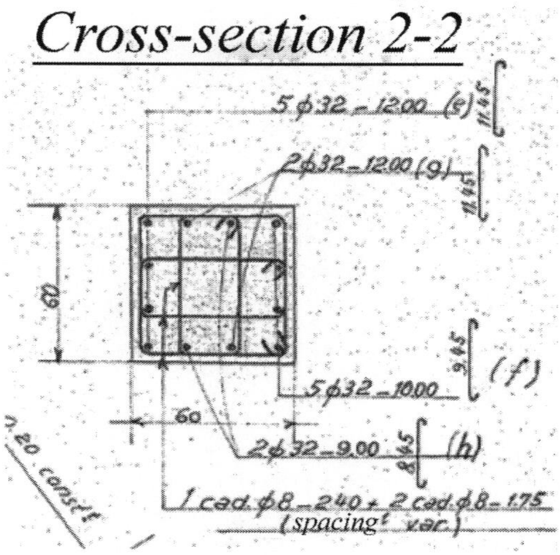

Figure 5.10. *Reinforcement details*

The formula to calculate the compressive stress on the concrete was $R_b = N/\Omega$, or in this case, a concrete resistance of 7 MPa and an admissible force of 3.15 MN (315 t) were considered.

Article 12 of the Instruction validated the fact that elements whose slenderness was less than 20 and whose concrete compressive stress remained below the limit value were not checked for buckling. In our case, the value was $l/a = 27/0.60 = 45 \geq 20$. Buckling must therefore be considered.

Typically, Rankine's formula was used:

$$N/\Omega\ (1 + kl^2/10\,000r^2) \leq R_b$$

with:

– N: total compressive force;

– l: length of the part;

– r: radius of gyration;

– k: coefficient depending on the conditions at the ends =1 (birotated bar).

Therefore, N = 2.79 MN (279 t).

If we assume that the brace works in compression and that the post works in tension, the capacity of the latter is given by N = 140 × 10 × 8.01 10^{-4} = **1.12 MN (112 t)**.

Figure 5.11. *Support posts. For a color version of this figure, see www.iste.co.uk/lauzin/concrete.zip*

Figure 5.12. *Stabilizing brace. For a color version of this figure, see www.iste.co.uk/lauzin/concrete.zip*

6

BA 45 Regulations

6.1. Overview

During 1944, the French Ministry of Reconstruction set up a commission chaired by Mr. Caquot to establish regulations for the use of reinforced concrete and steel applicable to buildings, with those of 1906 and 1934 applying mainly to bridges. The minister's introductory letter specified what was taken into account:

> [...] for the construction of reinforced concrete elements of public buildings, the instructions annexed to the circulars of October 20, 1906, and July 19, 1934 of the Ministry of Public Works were followed, even when these buildings were not truly civil engineering structures, which the instructions were more specifically tailored to.

The purpose of these regulations is summarized in the foreword:

> The application of these new guidelines will allow lower expenses than the requirements respected until now by good manufacturers.

These BA 45 regulations included nine sections as follows:

– section 0: forward;

– section 1: nature and quality of materials;

– section 2: admissible constraints;

– section 3: resistance calculations;

– section 4: presentation of projects;

– section 5: execution of works;

– section 6: work tests;

– section 7: appendix 1;

– section 8: appendix 2.

6.2. Calculation bases: requirements

6.2.1. *Basic requirements*

The following principles were considered:

– the "admissible stresses" were calculated;

– the actions were taken into account with their characteristic values.

6.2.2. *Project usage duration*

The project usage periods were not mentioned in the Instruction.

6.2.3. *Thermal effects*

The coefficient of thermal expansion was set at 1.1×10^{-5} m/m/°C. The variations to be taken into account in the calculations were:

– ±20°C in Metropolitan France;

– ±30°C in French Overseas Territories.

Except for the calculation of arches and similar hyperstatic structures, the effects of shrinkage and temperature were not introduced into the calculations from the moment when the distance between joints was at a maximum of:

– 25 m in dry regions or regions with strong temperature differences (Mediterranean region);

– 50 m in humid and temperate regions.

The shrinkage coefficient was 2×10^{-4} m/m above the Bayonne–Grenoble line and 3×10^{-4} m/m below. In North Africa, it was 4×10^{-4} m/m.

6.2.4. *Differential settlements and movements*

There are no specific data on this subject in the Instruction.

6.3. Materials

6.3.1. *Concrete*

Concrete could be composed of:

– 250–315 class cement (artificial Portland cement and iron (slag) cement) without specific recommendations);

– class 160–250 cement is subject to control tests or for low-difficulty structures in which the constraints of the materials are low.

Other classes of cement may be admitted upon full justification.

The minimum cement dosage was 350 kg/m^3. The maximum diameter of the largest aggregate was set at 25 mm.

6.3.1.1. *Strength*

Article 2.2 of the Instruction limited the value of the mechanical resistance of concrete to compression to:

– $R_b = 28/100 \times$ 90-day compressive strength for cast-in-place concrete;

– $R_b = 33/100 \times$ 90-day compressive strength for precast elements in workshop with permanent monitoring.

The allowable tensile stress was equal to $R'_b = 28/100 \times$ 90-day tensile strength.

Suitability tests ("preliminary tests") were required before any construction. These tests were carried out on test specimens at 2, 7, 28 and 90 days. It was permitted for concrete formulated with cement of classes 160–250 and 250–315 that:

– compressive stress $n_7 = 0.57\ n_{90}$ and $n_{28} = 0.83\ n_{90}$ (where n_j denotes the compressive stress at j days);

– tensile stress $n'_7 = 0.70\ n'_{90}$ and $n'_{28} = 0.90\ n'_{90}$ (where n'_j denotes the tensile stress at j days).

Without testing, the strength values of the concretes formulated with class 160–250 cements cast in place were:

– $R_b = 50$ kg/cm^2 and $R'_b = 5$ kg/cm^2 for a cement content of 350 kg/m^3;

– $R_b = 56$ kg/cm^2 and $R'_b = 5.6$ kg/cm^2 for a cement content of 400 kg/m^3.

For concretes formulated with class 250–315 cements, these values became:

– $R_b = 63$ kg/cm² and $R'_b = 6.6$ kg/cm² for a cement content of 350 kg/m³;

– $R_b = 72$ kg/cm² and $R'_b = 7.2$ kg/cm² for a cement content of 400 kg/m³.

Higher values were allowed for cast-in-place or prefabricated structures subject to inspection (approximately 10% increase).

The internal friction angle of the concrete was 37°.

6.3.1.2. *Elastic strain*

The elastic modulus E (bars) of the concrete was calculated from the following formulas:

– $E_v = 6{,}000\ (n_j)^{0.5}$ under a combination of permanent and variable loads;

– $E_i = 18{,}000\ (n_j)^{0.5}$ under rapidly variable loads (application period of less than 24 h).

Poisson's ratio was 0.15.

The steel/concrete equivalence coefficient was 15 when the concrete used was based on class 160–250 cements.

6.3.2. *Reinforced concrete steels*

6.3.2.1. *Properties*

The steels commonly used in reinforced concrete constructions had a breaking strength of approximately 40–50 kg/mm². The elastic modulus of the steel was 21,000 kg/mm².

6.3.2.2. *Resistance*

In case the *concrete cracking was considered minimally damaging* (BAEL), the tensile stress of steel R'_a was:

– $R'_a = 0.6 \times n'_e$ for values of $n'_e \leq 24$ kg/mm²;

– $R'_a = 0.4 \times n'_e + 4.8$ for values of $n'_e \geq 24$ kg/mm².

Here, n'_e is the elastic limit of steel.

In the case of class 160–250 cement, the admissible tensile stress was limited to 12 kg/mm².

If the *concrete cracking was considered damaging* (BAEL), the tensile stress of the steel R'_a was:

– $R'_a = 14.4$ kg/mm² for concrete with 350 kg/m³ cement and a resistance greater than 280 kg/cm² at 90 days;

– $R'_a = 16$ kg/mm² for concrete with 400 kg/m³ cement and a resistance greater than 320 kg/cm² at 90 days.

NOTE.– The cases mentioned where cracking was considered detrimental were mainly structural parts subjected to weathering.

NOTE.– In the case where the structure was in permanent contact with water on one side (reservoir, etc.), the value of R'_a was $R'_a = 10$ kg/mm²; otherwise (foundations, for example), the value of R'_a was $R'_a = 12$ kg/mm².

6.3.2.3. *Steel/concrete adhesion*

Article 2.3 of the Instruction sets the value of the longitudinal sliding on itself and its adhesion to the metal at:

$$R_d = 2R'_b/(1 + \Phi/d_1)\,(1 + \Phi/d_2)$$

where R_d is the “admissible adhesion” constraint, Φ is the diameter of the bar, d_1 is the minimum distance from the axis of the bar to the free surface of the concrete, d_2 is the smallest distance from the axis of the bar to the free surface of the concrete in both perpendicular directions and R'_b is the tensile strength of the concrete.

The straight sealing length was then defined as $\mathbf{l_s = \Phi/4 \times R'_a/R_d}$.

The overlap length of the tensioned bars was equal to the straight sealing length l_s. If the bars had normal hooks and if $R'_a/R_b = 20$, then for coated bars of diameter Φ and at least 3Φ apart between axes, the overlap length was $0.60\ l_s$.

The tensioned bars should not overlap by more than 50% in the same section.

6.4. Durability and cover of reinforcements

6.4.1. *Overview*

The durability criteria are not mentioned in the Instruction, nor are the conditions of exposure of the works to physicochemical attacks.

6.4.2. *Durability*

Service lifetimes are not mentioned.

6.4.3. *Cover*

The covers were set to the following values:

– 1 cm for structures where one of the facings is exposed to weathering;

– 2 cm for structures where both facings are exposed to weathering.

6.5. Structural analysis

6.5.1. *Overview*

The purpose of structural analysis is to determine the distribution of stresses or constraints and strain. The behavior model used in the Instruction is that of the *linear elastic behavior*.

Article 3 states:

> The calculations will be conducted by scientific methods supported by experimental data; however, in certain specific cases, direct experiments on a typical element, carried out in the elastic domain and then pushed to breaking, will be considered as complete justification of the elements.

6.5.2. *Load combinations*

Article 3.2 of the Instruction requested that the following be considered in the resistance calculations:

– permanent loads;

– of the operating or test load, which includes climatic loads;

– thermal efforts (for non-freely expandable structures);

– effects of shrinkage;

– the construction method.

These loads were combined without load factors and considered the maximum value.

NOTE.– Typically, the density of reinforced concrete is 2.5 t/m^3, and the formula is identical to that of the previous rules.

The operating and climate loads were set according to standards.

6.5.3. *Second-order effects*

Second-order effects that can significantly affect the *overall stability* of the structure were not considered in the Instruction.

6.5.4. *Modeling the structure*

The regulations did not include digital modeling of structures.

6.6. Structural calculations

6.6.1. *Simple compression*

Note that for the forces applied to the posts, the value of the vertical loads was calculated by considering the absence of continuity of the beams at the top of the post except in the case of structures with two continuous spans where the connection between the post and the beams was taken into account.

These vertical loads were increased by:

– 10% for posts receiving two continuous beams or four cross beams on each floor;

– 15% for posts receiving two continuous beams on each floor but each of which only supports a single floor span or a single beam supporting two floor spans or three beams forming a T;

– 20% for posts supporting on each floor a single beam loaded by a single floor span or two perpendicular beams.

For the posts participating in the bracing of the building, it was accepted that the latter were embedded at the level of each of the floors and articulated at mid-height of each floor and either articulated or embedded at the level of the foundations.

The minimum reinforcement percentages are those indicated in Table 6.1.

Compressive stress	Minimum percentage of longitudinal reinforcements	Minimum percentage of transverse reinforcements
Greater than 0.75 × allowable stress	0.6%	0.3%
Other cases	0.6% × S_b	0.3% × S_b

Table 6.1. *Minimum percentages of reinforcements*

NOTE.– S_b is the surface of the column, which would undergo a compressive stress equal to 0.75 × allowable stress.

Section 3.513 dealt with confined concrete and in particular gave a formula for calculating the increase in the mechanical resistance of concrete in compression provided by confinement.

6.6.2. *Compression with bending*

The calculation was carried out for columns whose buckling modulus $\lambda \geq 0.5$ (see section 6.6.4), checking that:

– under normal stress alone, the permissible stresses of the materials were not exceeded;

– in compound bending, under the action of the normal force and the bending moment increased by the moment amplification factor $N_c/(N_c - N)$ (N being the normal effort and N_c the Euler critical load), the admissible stresses were not exceeded.

The constraints were then given by the classical formulas for the resistance of materials, namely:

– *concrete compression on the extreme fibers*:

$R_b = N/\Omega + (M\ v_b)/I$ with v_b distance from the extreme fiber to the axis of inertia

The formula was modified to take into account the slenderness through the radius of gyration r ($I = r^2\ \Omega$) according to:

$$R_b = N/\Omega\ (1 +/- (v_0 v_b/r^2))$$

where v_0 is the eccentricity of N relative to the geometric axis;

– *compression of steels on extreme fibers*:

$R_a = m\,(N/\Omega + (M\,v_a)/I)$ with v_a distance from the extreme steel to the axis of inertia

6.6.3. *Bending*

Calculation assumptions:

– the beams are assumed to be horizontal and subjected to downward vertical forces;

– the free span (except for beams on bearing devices) is the clear distance between the supports;

– in the elements that are continuous or integrated into the supports, the sections to be checked are those located on the inside face of the supports.

6.6.3.1. *Case of continuous beams*

If $h \leq l/2$ (h beam height, l free span), the support moments of the continuous beams were calculated either by applying the resistance of the materials or by the following simplified methods:

– The Caquot method (see Figure 6.1):

— 1 A uniformly distributed load p_s and p_d per unit length, will give a support moment equal in absolute value to

$$\frac{p_g l'^3_g + p_d l'^3_d}{8{,}5\,(l'_g + l'_d)}$$

— 2 a concentrated load Ps or Pd at a distance a from the support surface will give a support moment equal in absolute value to

$$\frac{kP_g l'^2_g}{l'_g + l'_d}\left(\text{or } \frac{kP_d l'^2_d}{l'_g + l'_d}\right)$$

Figure 6.1. *Values of moments on supports according to the Caquot method*

The values of k and k′ are given by the scale in Figure 6.2.

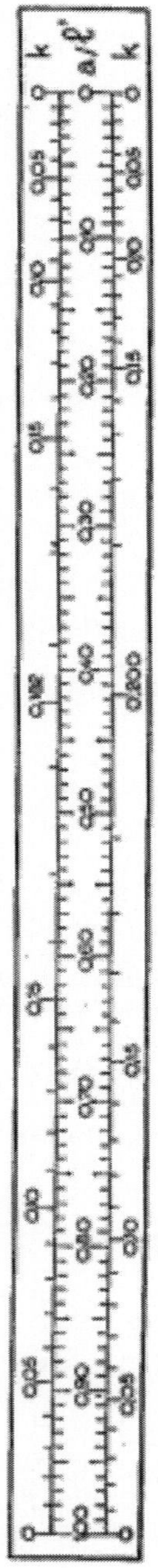

Figure 6.2. *Values of the coefficients k and k'*

Beam type	Support condition	Span moment	Edge support moment	Central support moment/central span moment	Support moment adjacent to the edge spans
A single span	Partial embedding on the two supports	0.80 M_0	–0.40 M_0	0	–
	Partial embedding on a single support	0.90 M_0	–0.40 M_0	0	–
Two spans (whose free spans do not vary by more than 20%)	Partial embedding on the edge supports	0.70 M_0	–0.25 M_0	–0.70 M_0	–
	Free edge supports	0.75 M_0	–	–0.80 M_0	–
More than two spans (whose free spans do not vary by more than 20%)	Partial embedding on the edge supports	0.75 M_0	–0.25 M_0	–0.50 M_0/0.66 M_0	–0.60 M_0
	Free edge supports	0.80 M_0	0	–0.50 M_0/0.66 M_0	–0.66 M_0

Table 6.2. *Values of support and in span moments*

Span moments were then calculated using the *material resistance* formulas.

NOTE.– In these formulas, l′ = l at the edge spans l′ = 0.8l for a continuity span.

– Flat rate method:

The conditions for applying the flat rate method are given in section 3.521 (11.12.13). For example:

– structural posts are not supported by floors;

– the variable loads applied to the floors are less than 500 kg/m² distributed and 1,000 kg punctually;

– variable loads are less than 3/5 × total load or 3/2 × permanent load.

The support and span moments are summarized in Table 6.2.

NOTE.– M_0 is the moment of the associated isostatic span.

6.6.3.2. *Case of partition beams*

Bent elements for which the section height exceeded five-tenths of the free span between the internal faces of the supports were considered partition beams; they were calculated accordingly. In particular, the calculation rules and reinforcement arrangements given in Appendix 7.7 of the BA 45 Regulations could be used in the following two cases:

1) when a beam had a height between the half-span and the span, the lower part of this beam, with a height equal to the half-span, was considered an active element and established in accordance with the rules of the first case, and the remaining part, which surmounts it, was established according to the circumstances based on the role played by the structure overall;

2) when a beam had a height greater than the span, the lower part of this beam, which is equal in height to the span, was considered as an active element and established in accordance with the rules of the second case, and the remaining part, which surmounts it, was established according to the circumstances according to the role played by the overall structure. It was also possible, in certain cases, to consider the partition beam as an active element only when the lower part had a height equal to half the span, which was then assessed according to the rules of the first case.

6.6.3.3. *Case of isostatic beams*

The bent beams could be considered with or without support embedding. In the latter case, two possibilities exist.

1) Partial embedding was possible on both supports.

Span moment +0.80 M_0

Support moment –0.40 M_0

Proof that each of the elements serving as support could resist a moment of 0.40 M . The shear forces were equal to those of the comparison span.

2) Partial embedding was only possible on a single support.

Span moment +0.90 M_0

Support moment –0.40 M_0

The shear forces were then calculated taking into account the support moment 0.40 M_0.

6.6.4. *Buckling*

Article 3.514 of the regulations took into account the following assumptions:

– *theoretical height of a post h*: distance between the upper faces of two consecutive floors or between the top of the foundation and the top of the upper basement floor;

– *buckling length l_f*: the values are summarized in Table 6.3.

End conditions	l_f value	Observations
One support is embedded, the other is free	2h	
One support is embedded, the other is articulated	h/√2	
Both articulated supports	H	
Both embedded supports	H	If the two ends are not blocked in movement relative to each other
Both embedded supports	h/2	If the two ends are blocked in movement relative to each other
For multi-story buildings	0.7h or 0.9h	0.7h if the post is supported on a foundation block or on beams with at least the same inertia as it; otherwise, the value is 0.9h

Table 6.3. *Buckling length values*

Article 3 of the rules then defined a *unit of buckling f* such that $f = 4r\sqrt{E_b/n_j}$ as well as a buckling modulus $\lambda = l_f/f$.

In this formula:

– r refers to the radius of gyration;

– E_b is the elastic modulus of the concrete;

– l_f is the buckling length.

Depending on the values of λ obtained, the loads were multiplied by a buckling coefficient. This coefficient took the values indicated in Table 6.4.

Values of λ	Buckling coefficient values
≤ 0.5	Buckling not taken into account
$0.5 \leq \lambda \leq 1$	2λ
$1 \leq \lambda \leq 1.5$	$1 + \lambda^2$
$\lambda \geq 1.5$	$1.44\lambda^2$

Table 6.4. *Buckling coefficient values*

6.7. Constructive provisions

The main provisions related to the execution of the works are described in Chapter 5 of the Instruction under the title "Execution of the works".

6.7.1. *Pouring concrete*

No specific instructions were given, except that in the case of sealing the concrete with mechanical methods, the test pieces had to be implemented in the same way. The reinforcements had to be clean without traces of rust or grease.

6.7.2. *Distance between reinforcements and between reinforcements and the formwork*

The distance between the steel as well as between the steel and the formwork had to allow any voids to be filled with concrete.

– For structures where no facing was exposed to the elements, the cover was set at 10 mm of concrete.

– A minimum cover of 20 mm to meet the durability criterion was required for facings in contact with weather.

– For aggressive atmospheres, the cover was fixed by the specifications.

6.7.3. *Construction joint treatment*

The construction joints were required to be perpendicular to the direction of the compression stresses. In each construction joint, the following steps must be performed:

– "raw" cleaning of old concrete;

– wetting of the recovery surface;

– fresh concrete was poured.

NOTE.– The use of cement slurry for replacing concrete was not recommended.

On the other hand, the cement dosage was increased in the first layer of new concrete in contact with existing concrete, with the reduced size of the aggregates.

6.7.4. *Concrete curing*

The formwork and concrete were moistened until they fully hardened. Watering of fresh concrete was permitted if it did not damage the surface of the structure.

6.7.5. *Concrete inspection tests*

The tests were carried out on batches of at least three test samples. These test pieces consisted of cubes of:

– 100 mm on each side if the size D of the aggregate was less than 25 mm;

– 141 mm on each side if the size D of the aggregate was between 25 mm and 50 mm;

– 200 mm on each side if the size D of the aggregate was greater than 50 mm.

6.8. Application to the calculation of circular tanks

6.8.1. *Calculation of the main reinforcements*

The manual calculation of the forces in the walls of the circular tanks was carried out by considering horizontal parts of the tank wall of 1 m in height. The tensile force in the hoops was then determined in each slice by the formula $T = \omega Rzh$, in which:

– ω is the density of the liquid;

– R is the radius of the tank;

– z is the depth to the center of the layer at the liquid level;

– h is the total height of the wall.

For example, for a wall 5m high the diagram of the forces was considered as indicated in Figure 6.3.

For layer AB, for example, considering a radius of 4 m for the tank and water as content, the tensile force in the hoops was (h is then the height of the layer):

$$T = \omega Rzh = 1 \times 4 \times 2.50 \times 1 = 10 \text{ t}$$

NOTE.– This 1 m layer height was sometimes considered too great; in fact, if we calculate at point A, the traction value is 8 t; on the other hand, at point B, it is 12 t, which is higher than that calculated.

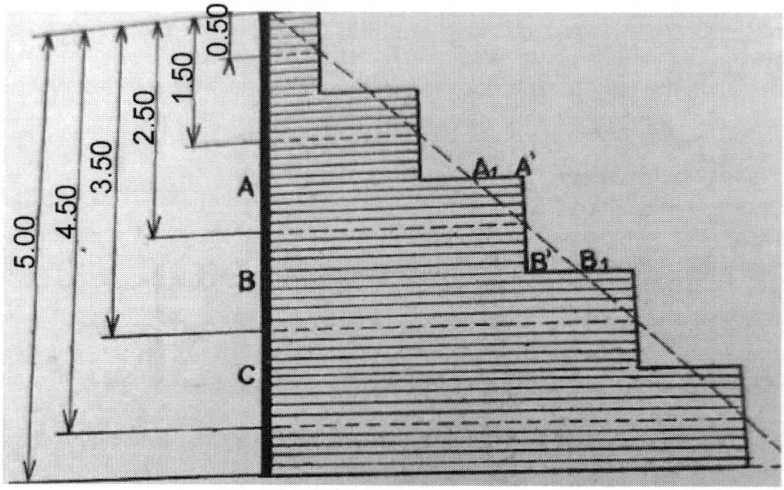

Figure 6.3. *Diagram of the forces on the wall. For a color version of this figure, see www.iste.co.uk/lauzin/concrete.zip*

As a result, some engineering offices have calculated layer heights at 0.50 m.

The steel section A_s intended to balance this tensile force was then given by:

$$A_s \times R_a = \omega Rzh$$

In this formula, R_a is the admissible tensile stress of the steels, that is, $\mathbf{R_a = 1{,}000}$ *bar* (100 MPa).

6.8.2. *Calculating the distribution reinforcement*

The BA 45 regulations specified the distribution reinforcements in the floor framework supporting distributed loads. The recommended section was then equal to a quarter of that of the main reinforcements.

In the case of a cylindrical tank, the role of the distribution reinforcements was to take up the parasitic forces due to thermal expansion, shrinkage, etc. For this reason, a standard steel section equal to *one-third of the section* of the main reinforcements is recommended.

The following principles were generally adopted:

– For small-diameter tanks, a distribution reinforcement of a constant section equal to one-third of the section of the main reinforcements was used over the entire wall. The spacing was between 20 and 25 cm.

– For medium-diameter tanks, the bottom of the wall was reinforced with tightly distributed reinforcements (spacing between 10 and 12 cm). The section of these reinforcements was equal to one-third of the lower main hoops. From the mid-height of the wall, the spacing of the secondary steels was then increased (20–24 cm).

– For large tanks, the diameter of the distribution reinforcements is often adapted to that of the main reinforcements in compliance with the one-third rule.

6.8.3. *Calculation of concrete wall thicknesses*

The wall was sized to withstand:

– the thrust of the liquid n_b;

– the shrinkage effects R_b.

The forces caused by the thermal gradient, expansion, clamping, etc., were generally not considered. It was then verified that:

$$R_b + n_b \leq \sigma_b$$

where $\boldsymbol{\sigma_b}$ is the rate of work of the concrete in tension, that is, in the most frequent cases, it has a value of approximately 35 bar (3.5 MPa).

The value of the shortening due to concrete shrinkage was considered equal to 2×10^{-4} m/m.

The value of the thickness of the concrete wall was obtained by calculating the tensile forces in a layer dz of the cross-section, that is,

– a tensile force of the steel equal to n_a A_s dz, where n_a is the tensile stress of the steels and A_s steel section per lm;

– a concrete tensile force equal to n_b (e + m A_s) dz, where n_b is the tensile stress of the concrete and of the section of the concrete per lm, m is the steel–concrete equivalence coefficient.

By equalizing these values, the thickness e of the plate is $e = A_s (n_a/n_b - m)$.

At a depth of z, the thrust of water of density ρ associated with a circular tank of radius R was balanced by the steel section A_s, hence:

$$e = \rho R z (k - m)/n_a$$

The limit value imposed at the top of the wall was then 6–8 cm. From this, an approximate formula was deduced:

$$e = 0.0035Rz$$

NOTE.– It was recommended in the literature to limit the compressive stress of concrete to 20 bar (2 MPa instead of 50 bar in the 1945 regulation). Some manufacturers even used lower stresses of approximately 15 bar.

EXAMPLE 6.1.– For a 6-m high wall and a tank radius of 10 m, we obtained a wall with a variable section ranging from 21 cm at the base to 8 cm at the top of the wall, with a value at the mid-height of the wall of 10.5 cm.

Notably, these values were particularly low, did not consider the possibility of correctly concreting the wall and required meticulousness in the wedging of the steels to respect the covers.

EXAMPLE 6.2.– This application concerns the calculation of a tank on a tower. It is a 500 m^3 tank with an internal access chimney to the cover dome. The tank rests on a 6 m diameter tower.

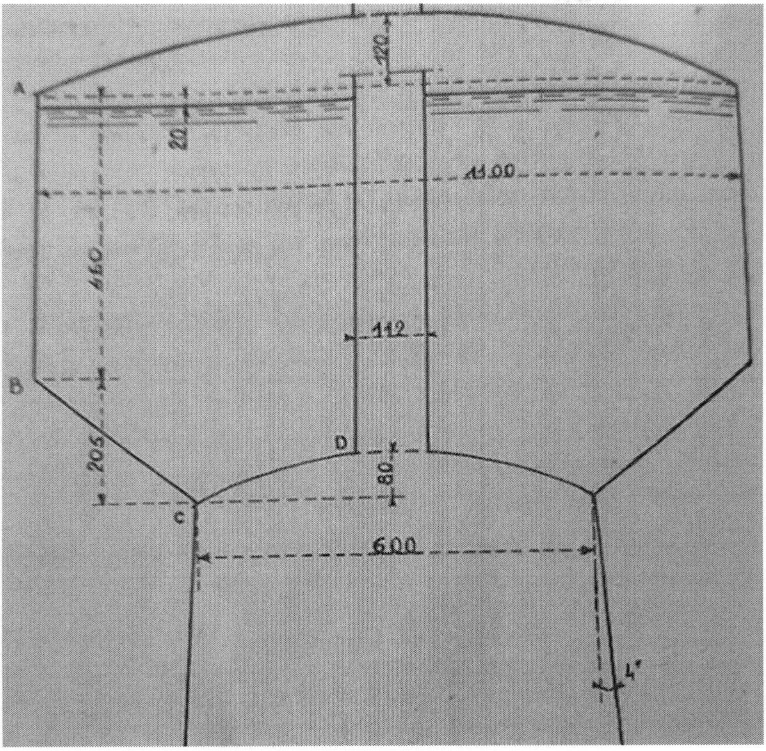

Figure 6.4. *Tank cross-section. For a color version of this figure, see www.iste.co.uk/lauzin/concrete.zip*

6.8.4. *Cover dome study*

6.8.4.1. *Verifying the thickness of the dome*

The thickness e of the lowered dome was defined by the compressive stress on the concrete obtained by dividing the normal force in the vault by 100 × e, that is, values of e greater than 6 or minimum 8 cm.

The arch deflection is given by the value D/10, that is, 1,100/10 = 1.10 m. A deflection of 1.20 m was chosen. The loads applied to the shell are then as follows:

– permanent loads:

- 6-cm-thick reinforced concrete dome: 150 kg/m^2,

- waterproofing cover on the dome: 0.025 × 2,000 = 50 kg/m^2;

– operating load: according to the standard, it was set equal to 100 kg/m².

NOTE.– This load was not concomitant with the snow load.

Let there be a total load of 380 kg/m².

The compression in the direction parallel to the covering dome is given by (see below):

– $\alpha_1 = 0$ and $\alpha_2 = 90°$;

– ρ_l: density of the liquid = 0;

– p: total load in m² (in replacement of ρ_b*e);

– f: dome spire;

– r: radius of the dome:

$$\sigma^*e = p\,(r^4 - 4r^2f^2 - f^4)/(4r^2f)$$

In our case:

$$\sigma^*e = 380\,(5.50^4 - 4 \times 5.50^2 \times 1.20 - 1.20^4)/(4 \times 1.20 \times 5.50^2) = 1{,}930 \text{ kg}$$

By limiting the compressive stress of the concrete to 20 bar, the section of the concrete of the dome would be 1,930/20 = 96 cm² per lm, or a thickness of 0.96 cm. The minimum value of 6 cm was then retained. The minimum reinforcement of the dome was then made up of a grid of steels *6 mm in diameter and spaced 15 cm apart*.

6.8.4.2. *Verifying the traction in the lower belt*

The tensile force is:

$$T = p\,(r^4 - f^4)/(4rf)$$

In our case:

$$T = 380 \times (5.50^4 - 1.20^4)/(4 \times 5.50 \times 1.20) = 13{,}140 \text{ kg}$$

The steel section was obtained by considering a steel working rate of 1,000 bar, that is, $\mathbf{A_s}$ **= 13.14 cm² (5Φ20)**. The concrete section was then equal to 0.035 × T = 460 cm² (30 × 16, for example).

6.8.5. *Vertical wall study*

The geometric assumptions are as follows:

– tank radius: 5.50 m;

– water height in service: 4.40 m.

The tensile stress in the steel is 1,000 bar. The thickness of the tank wall was **e = 0.0035Rz**, which in our case is e = 0.0035 × 5.50 × 4.40 = 0.09 m. *The thickness chosen was therefore 9 cm* (with the possibility of decreasing this thickness to 8 cm in the upper part).

The steel sections needed to balance the liquid thrust were calculated by horizontal sections. Thus, for the section between –4.40 m and –3.40 m, for example, considering a radius of 5.50 m for the tank and water as content, the tensile force in the hoops was (h is the height of the section):

$$T = \omega Rzh = 1 \times 5.50 \times 3.90 \times 1 = 21.5 \text{ t}$$

The steel section required was therefore 21,500/1,000 = 21.5 cm² (7Φ20, for example).

The spacing of the steels was given by the formula: $dz = (\pi D^2 R'_a)/(4\omega Rz)$, where R'_a = 1,000 bar, ω = 0.001, R = 5.50 m, z = depth of the section and D = diameter of the steel. For the previously calculated section, dz = 785,000 × 2^2/(550 × 390) = 14 cm. At 3-m depth, the spacing then becomes: dz = 19 cm.

Overall, the calculation led to a total steel section of 51 cm² distributed over the height of the tank.

The distribution reinforcements were estimated at a flat rate of one-third of the section of the main reinforcements, that is, 10 mm diameter steel spaced approximately 12 cm apart in the lower half of the wall and then 10 mm diameter steel every 24 cm in the upper half.

The lower belt of the tank wall, which provides the connection between the vertical part of the wall and the truncated part of the skirt, was sized to take the following loads:

– cover dome: 1,095 kg/m;

– gutter, upper belt: 250 kg/m;

– tank wall with cover: 1,205 kg/m;

– thermal insulation wall: 620 kg/m;

– self-weight of the lower belt: 200 kg/m.

That is, the total load is 3,370 kg/m.

This load causes tension in the lower belt of P (R – r) × R/E = 3,370 × (5.50 – 3.00) × 5.50/2.05 = 22.6 t.

At point B, the water pressure is not considered since it has been included in the reinforcement of the wall.

The steel section required to balance this tension is: 22,000/1,000 = **22.60 cm²** (4Φ20 + 2Φ25, for example).

The concrete section of the belt is given by: 0.035 T = 0.035 × 22,600 = 790 cm² (35 × 25 cm², for example).

6.8.6. *Truncated cone study*

6.8.6.1. *Reinforcement of the truncated cone*

The wall thickness was given by the following formula:

$$e = x\ \omega\ \gamma/(\sin\alpha\ (R_b\, R'_a/(R'_a - m\ R_b) - \delta x/\mathrm{tg}\alpha))$$

with:

– α: inclination of the truncated cone;

– R_b: compressive strength of the concrete;

– R'_a: tensile strength of steel;

– m: steel/concrete equivalence coefficient;

– δ: density of reinforced concrete (2,500 kg/m^3);

– ω: density of the contained liquid;

– x and γ are defined in Figure 6.5.

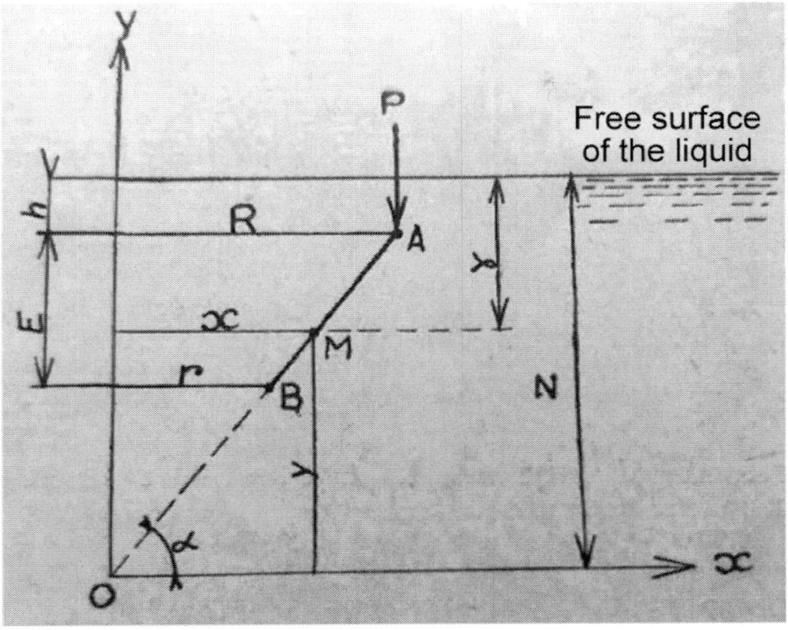

Figure 6.5. *Diagram of the truncated cone. For a color version of this figure, see www.iste.co.uk/lauzin/concrete.zip*

This formula leads to a plate thickness of *14.15 cm* at the high point of the truncated cone (B) and *11 cm* at the low point (C). The total traction in the truncated cone is given by the following formula:

$$T = (R^2 - r^2) \times (\omega\, Z + \delta\, e \cos\alpha)/(\sin 2\alpha) - (R^3 - r^3) \times \omega/(3\cos^2\alpha)$$

with $Z = h + E + r\ \mathrm{tg}\alpha = 4.40 + 2.05 + 3 \times 0.82 = 8.91$ m ($\alpha = 39°$). Therefore, $T = 198{,}655 - 77{,}750 = 121{,}000$ kg.

The required steel section is therefore 121,000/1,000 = **121 cm² (39Φ20**, for example).

This steel is distributed over a 3.23 m length of the truncated cone, that is, steel spacing between 8 and 10 cm.

The regulations did not impose the distribution of steel on one or two faces; in our case, it is preferable to implement them on both faces of the wall, which allows the spacing to be reduced to approximately 15 cm (ease of concreting and vibration).

The distribution steel was sized to one-third of the section of the load-bearing steel, that is, 39/3 = 13 cm² (Φ 16 steel approximately every 17 cm).

6.8.6.2. *Verifying the compressive stress in the truncated cone*

The formula for calculating the force at the meridians is given by:

$$F_{ec} = 1/r \sin\alpha \, [(\delta\, e/\cos\alpha + \omega\, Z)\,(R^2 - r^2)/2 - \omega\, tg\alpha\,(R^2 - r^2)/3 + PR]$$

Therefore, $F_{ec} = 1/(3 \times 0.634) \times ((2{,}500 \times 0.135)/0.773 + 8{,}910)\,(5.50^2 - 3^2)/2 - 820 \times (5.50^3 - 3^3)/3 + 3{,}370 \times 5.50 = 40{,}100$ kg.

The compressive stress in the 11-cm-thick concrete at the lowest point of the cone is:

$$R_b = 40{,}100/(100 \times 11) = 36.4 \text{ bar} \geq 35 \text{ bar}$$

A constant thickness of 14.5 cm is therefore maintained for the cone plate.

6.8.7. *Bottom dome study*

The same calculations as for the cover dome show that:

– the thickness of the vault can be 8 cm;

– the frames in the direction of the meridians: 3 cm²;

– distribution steel: 8 mm round steel every 20 cm.

The final reinforcement corresponded to the plan, as shown in Figure 6.6.

The steels of the lower belt of the bottom dome were determined by considering the diagram in Figure 6.7.

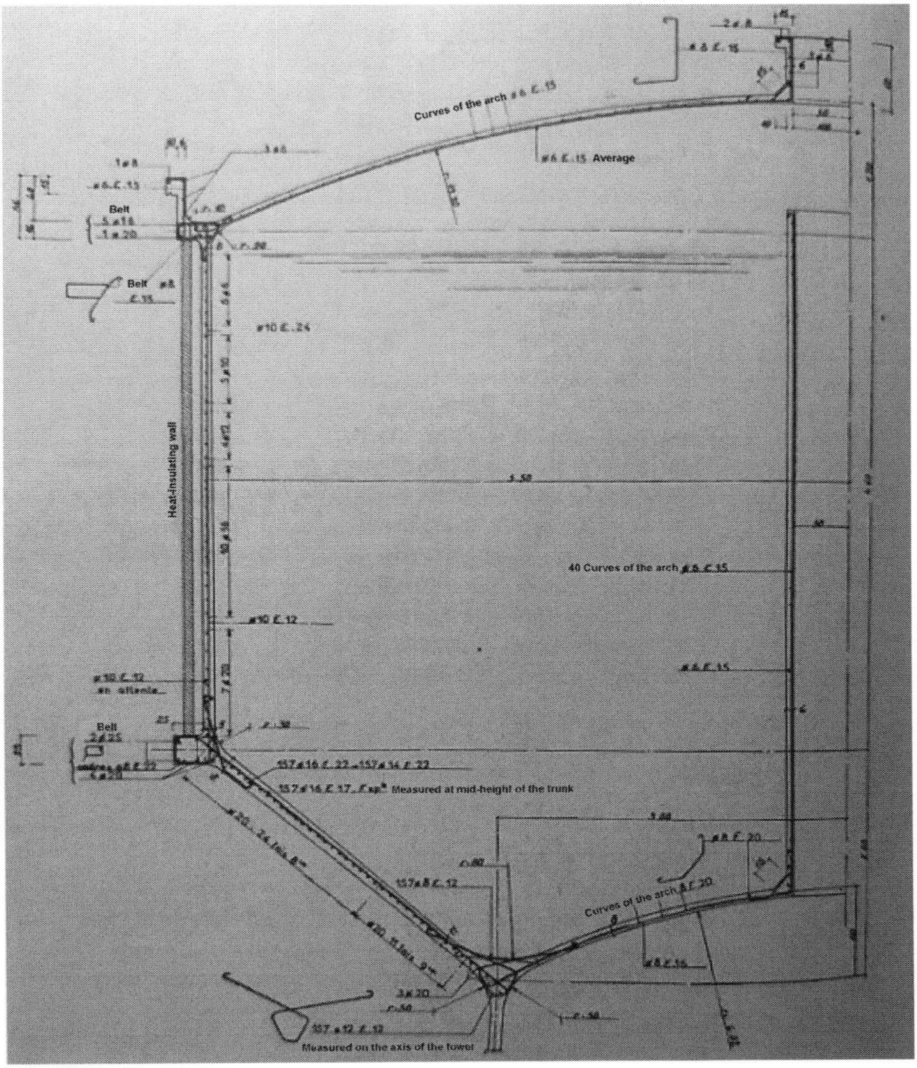

Figure 6.6. *Tank reinforcement plan. For a color version of this figure, see www.iste.co.uk/lauzin/concrete.zip*

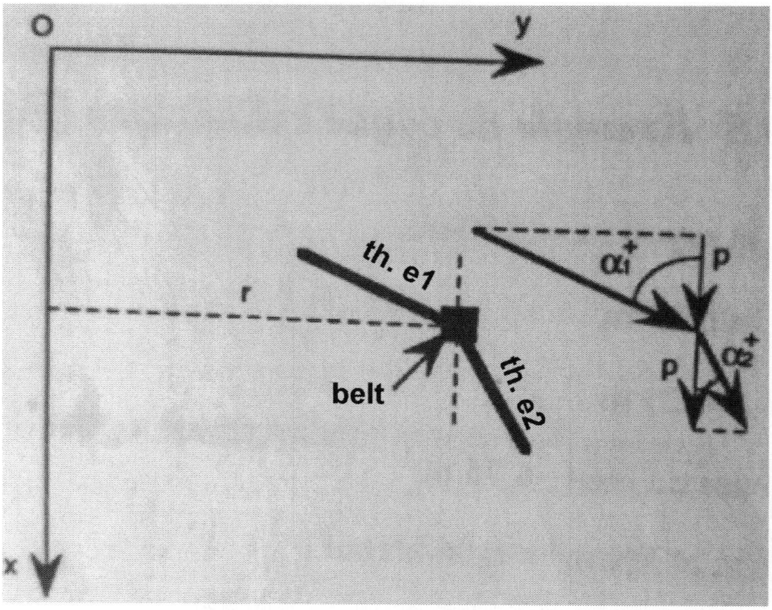

Figure 6.7. *Diagram of efforts at the lower belt level. For a color version of this figure, see www.iste.co.uk/lauzin/concrete.zip*

The force supported by the belt is therefore $T = p\ (tg\alpha_1 - tg\alpha_2)$, where p is the vertical distributed load.

In our case, the loads are as follows:

– vertical load full tank: $P_1 \sin\alpha = 40{,}100 \times 0.634 = 25{,}400$ kg;

– vertical load empty tank: $P_2 \sin\alpha = 12{,}000 \times 0.634 = 7{,}600$ kg;

– horizontal load full tank: $P_1 \cos\alpha = 40{,}100 \times 0.773 = 31{,}000$ kg;

– horizontal load empty tank: $P_2 \cos\alpha = 12{,}000 \times 0.773 = 9{,}300$ kg.

The bottom dome bears the loads:

– vertical load full tank: $P'_1 \sin\alpha = 18{,}700 \times 0.50 = 9{,}350$ kg;

– vertical load empty tank: $P'_2 \sin\alpha = 1{,}280 \times 0.50 = 640$ kg;

– horizontal load full tank: $P'_1 \cos\alpha = 18{,}700 \times 0.866 = 16{,}200$ kg;

– horizontal load empty tank: $P'_2 \cos\alpha = 1{,}280 \times 0.866 = 1{,}110$ kg.

The total load acting on the lower belt is then:

– vertical load full tank: $V_1 = 25{,}400 + 9{,}350 = 34{,}750$ kg;

– vertical load empty tank: $V_2 = 7{,}600 + 640 = 8{,}240$ kg;

– horizontal load full tank: $H_1 = 31{,}000 - 16{,}200 = 14{,}800$ kg;

– horizontal load empty tank: $H_2 = 9{,}300 - 1{,}110 = 8{,}190$ kg.

The angle between the tower and the vertical direction is 4°, and the previous formula therefore makes it possible to estimate the additional horizontal force linked to the non-verticality of the wall and acting on the lower belt, that is,

$$T = V\,(tg(4)) = 34{,}750 \times 0.07 = 2{,}420 \text{ kg}$$

The total thrust force on the lower belt is therefore $H = 14{,}800 + 2{,}420 = 17{,}220$ kg. The belt is therefore subjected to a compression force of $T = H \times r = 17{,}220 \times 3 = 51{,}660$ kg. This makes it possible to determine the squaring of the lower belt, that is, $S = 51{,}660/35 = 1{,}480$ cm².

It was not necessarily rectangular but rather trapezoidal in shape to integrate the anchors of the steel of the truncated cone and the bottom dome (75 cm base and 40 cm height, for example).

With the belt compressed both empty tanks and full tanks, it was not necessary to implement longitudinal reinforcements, only transverse reinforcements with a longitudinal bar.

NOTE.– With respect to the material used, it is possible to estimate the quantities at:

– 42 m³ for the concrete of the tank;

– 9 t steel for the same structure.

6.9. Practical considerations on tank design according to the 1945 regulations

6.9.1. *Tank compartmentalization*

The same tank could be used both for firefighting and as a drinking water reserve. It was customarily used to compartmentalize the tank: the fire reserve was in the outer portion, and the drinking water was in the inner portion. This arrangement made it possible to use the water from the fire reserve as thermal insulation and to keep the drinking water at a constant temperature.

6.9.2. *The tank cover*

Most drinking water tanks built at this time were closed, with the cover being either a slab or a dome. The latter solution was generally the least expensive because of the small amount of concrete and steel required to make it. It was also chosen because it was considered that rainwater drained away more quickly and that it was therefore not necessary to provide additional waterproofing.

Rainwater drainage remained a problem. Three possibilities were considered:

– lowering pipes from the outside (questionable aesthetics and difficult maintenance);

– implement drains (on the tower and the tank);

– send the water into the overflow tank. This was the solution that was chosen by the majority to provide a waterproof pipe without a joint.

6.9.3. *The external vertical walls of the tank*

Notably, some walls had horizontal cracks due to re-concreting. It was therefore recommended to pour the wall in one go. However, this was only possible in small tanks. For larger structures, it was necessary to prepare for re-concreting by cleaning, tapping and planning the concrete for the next phase with the same dosage and without layering thin cement. The overlaps of the distribution reinforcements in this area were offset.

6.9.4. *The interior vertical walls of the tank*

In the case of two-compartment tanks, the inner wall was sized in compression and tension. The problem often came from the lower part, where it was necessary to ensure the steel continuities of the different structures (truncated cone, bottom dome, etc.).

6.9.5. *Bottom dome*

To be able to construct a dome that was 8 cm thick, it was recommended to use concrete with an adequate grain size and a little dry to compact it well. The problem of wedging thin steel was often mentioned when pouring the concrete.

6.9.6. *Tank support*

For large structures, the tank support tower could play several roles:

– withstand the vertical forces provided by the tank as well as the wind forces;

– house living quarters or storage areas, workshops on each floor, etc.

Figure 6.8. *Example of a tank support. For a color version of this figure, see www.iste.co.uk/lauzin/concrete.zip*

For more modest works, the tower was made:

– either in the form of a braced pylon;

– or in the form of a tower.

This latter solution was often preferred for technical (passage of pipes, etc.) and budgetary reasons. The ideal solution was the truncated cone tower, which came closest to the support of equal resistance but was more delicate to construct.

The thickness of the towers was generally greater than 8 cm, but there were companies that went down to 6 cm (e.g. for a 1,000 t tank and a compressive stress on the concrete of 40 bar, it was possible to have a tower 15.20 m in diameter and 7 cm thick or a tower 5.30 m in diameter and 20 cm thick).

Circular NV 46 (the same snow and wind regulations of 1946) also required consideration of the constraints linked to the ovalization of the tower under wind loads.

With respect to the foundation of the tower, the geotechnical study was often carried out only for large structures; it consisted of performing one or more surveys under the foundation. The depth of the surveys was often stopped at a distance of three times the width of the latter.

The towers were mainly made of sliding or climbing formwork.

6.9.7. *General recommendations*

Figure 6.9. *Absence of water droplets in the upper part of the truncated cone. For a color version of this figure, see www.iste.co.uk/lauzin/concrete.zip*

Figure 6.10. *Thermal insulation of the tank by a brick wall. For a color version of this figure, see www.iste.co.uk/lauzin/concrete.zip*

Recommendations for the execution of structures were mentioned in the literature as follows:

– Provide dripstones at all overhangs: they were requested to have a large cross-section and to reinforce them with a large diameter bar (20 mm minimum). This arrangement was intended to avoid damage (see Figure 6.9).

– For two-compartment mushroom structures, the central section must be made in continuity with the tower.

– Frost precautions and insulation: various measures were taken to prevent the water from freezing in the tanks (floating compressible tower, a large inner tube, etc.). Insulation of the tank made it possible to overcome the problem of frost and heat. The truncated cone was rarely insulated due to the difficulty of construction. The tank was often insulated with bricks with sealing criteria to prevent water stagnation in the annular space (weep holes were provided). The insulation of the top of the tank was sometimes made of a double slab, hollow bricks or cellular concrete. Notably, even if it was not envisaged at that time, the insulation made it possible to take into account the effects of ovalization and cracking of the tank under asymmetrical thermal effects.

6.10. Practical considerations on the design of silos according to the 1945 regulations

6.10.1. *Overview*

The classic layout of a silo installation includes:

– a receiving tank into which the cereals are poured upon arrival;

– an elevator taking them to the floor above;

– a conveyor belt carrying the grains from the elevator to the cell filling port;

– the units making up the silo;

– a lower conveyor belt taking up the grains when emptying the cells.

With respect to the dimensioning of silo structures, the thrust and friction forces on the walls were often the subject of empirical calculations. Thus, until the 1940s, the first constructors sought to apply the theory of retaining structures to the calculation of silos, while other tank theories involved placing stress on walls with a liquid of the same density as the ensiled material. It was not until the experiments of Prante (Hamburg) or Airy (France) that these theories were considered erroneous and that the grain thrust and the pressure at the bottom of a cell did not increase indefinitely and instead tended toward limiting values. Some authors have proposed theories based on the assumption that the p/q ratio between the lateral thrust and the vertical pressure exerted by the grain is consistent.

In France, since 1943, the Technical Institute of Building and Public Works, under the direction of Mr. Reimbert, has carried out numerous tests on scale models and on full-scale silos. These tests were supplemented by those undertaken by the Union Chamber of Reinforced Cement Builders. A commission composed of representatives of rural engineering, builders, engineers from the Building and

Public Works Laboratory was tasked with establishing the test program and drafting a conclusion. These tests focused mainly on:

– determining the characteristics of the silage material: internal friction angle, friction angle on smooth concrete walls, rough concrete, smooth sheet metal, rough sheet metal and density;

– determining the pressure and thrust on the walls when filling the cell;

– determining overpressures during cell emptying.

Nature of material	Friction angle on smooth concrete	Friction angle on rough concrete	Natural slope angle	Apparent density (kg/m³)
Wheat	20°	25°	24°–26°	750–840
Oats	20°	25°	26°–28°	500–540
Barley	18°	24°	25°–29°	550–690
Corn	17°	23°	26°–29°	750–820
Millet	Between 25° and 28°	Between 22° and 28°	20°–24°	700
Rapeseed	Between 18° and 19°	Between 22° and 27°	27°	695
Turnip	Between 17° and 21°	Between 33° and 40°	28°–40°	650
Poppy	Between 22° and 24°	Between 32° and 35°	29°–30°	600
Flour	30°	40°	35°–45°	500–800
Cement	25°	29°	30°–40°	1,400–1,500
Ashes	36°	40°	35°–40°	635–750
Dry sand	27°	38°	38°–45°	1,600–1,800
Crushed stones	27°	38°	40°–45°	1,700–1,850

Table 6.5. *Friction angle values*

The values obtained are summarized in Table 6.5.

NOTE.– The natural angle of a slope was frequently equated to the internal friction angle.

The values from EN 1991.4 are presented in Table 6.6.

Material type	Specific weight γ (en kN/m³)		Angle of repose Φ_r (in degrees)	Internal friction angle Φ_i (in degrees)		Lateral stress coefficient K		Wall friction coefficient μ ($\mu = \tan \varphi_w$)				Localized load reference coefficient C_{op}
	γ_l	γ_u		φ_{im}	a_φ	K_m	a_K	type D1	type D2	type D3	a_μ	
	lower	upper		mean	factor	mean	factor	mean	mean	mean	factor	
Fly ash	8.0	15.0	41	35	1.16	0.46	1.20	0.51	0.62	0.72	1.07	0.5
Flour ○	6.5	7.0	45	42	1.06	0.36	1.11	0.24	0.33	0.48	1.16	0.6
Iron ore pellets	19.0	22.0	36	31	1.16	0.52	1.15	0.49	0.54	0.59	1.12	0.5
Hydrated lime	6.0	8.0	34	27	1.26	0.58	1.20	0.36	0.41	0.51	1.07	0.6
Limestone powder	11.0	13.0	36	30	1.22	0.54	1.20	0.41	0.51	0.56	1.07	0.5
Corn ○	7.0	8.0	35	31	1.14	0.53	1.14	0.22	0.36	0.53	1.24	0.9
Phosphate	16.0	22.0	34	29	1.18	0.56	1.15	0.39	0.49	0.54	1.12	0.5
Potato	6.0	8.0	34	30	1.12	0.54	1.11	0.33	0.38	0.48	1.16	0.5
Sand	14.0	16.0	39	36	1.09	0.45	1.11	0.38	0.48	0.57	1.16	0.4
Slag clinker	10.5	12.0	39	36	1.09	0.45	1.11	0.48	0.57	0.67	1.16	0.6
Soya grains	7.0	8.0	29	25	1.16	0.63	1.11	0.24	0.38	0.48	1.16	0.5
Sugar ○	8.0	9.5	38	32	1.19	0.50	1.20	0.46	0.51	0.56	1.07	0.4
Beet pellets	6.5	7.0	36	31	1.16	0.52	1.15	0.35	0.44	0.54	1.12	0.5
Wheat ○	7.5	9.0	34	30	1.12	0.54	1.11	0.24	0.38	0.57	1.16	0.5

Table 6.6. *Characteristic values of silage materials per EN 1991.4*

6.10.2. *Formulas determined by the 1943 commission*

During the tests carried out, the vertical pressure increased with the height of the silage grains, but owing to the friction of the material in contact with the walls, it was only a fraction of the total weight of this material.

When the height of the grains inside the cell is very large, the force balanced by friction is such that the vertical pressure hardly varies.

This pressure then tends toward a maximum, as shown in Figure 6.11.

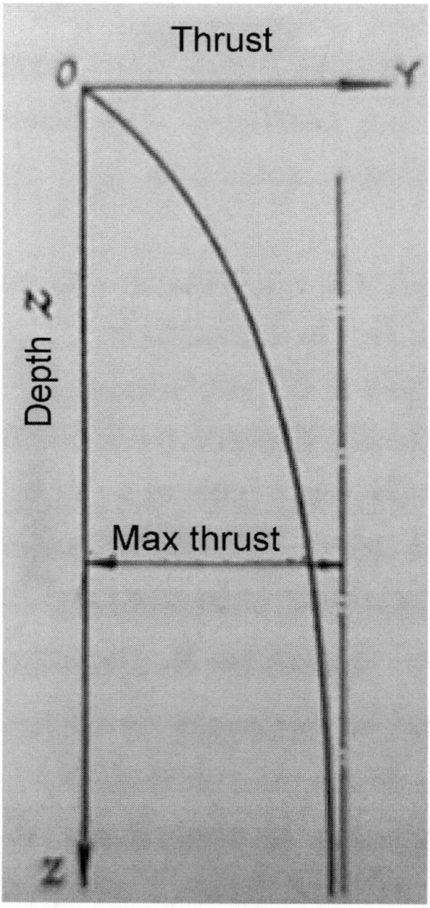

Figure 6.11. *Pressure and lateral thrust according to M. and A. Reimbert*

The experts then established general formulas for calculating silos based on their geometric shapes for *the filling phase*. These formulas are summarized in Table 6.7.

NOTE.– A, the characteristic abscissa, has the magnitude of a length and represents the ordinate at the origin of the asymptote to the curve of loads balanced by friction against the walls.

Cell type	Cylindrical	Regular polygonal base	Rectangular
Vertical pressure q_z	$q_z = \rho\,[z\,(z/A + 1)^{-1} + h/3]$	$q_z = \rho\,[z\,(z/A + 1)^{-1} + h/3]$	$q_z = \rho\,[z\,(z/A_2 + 1)^{-1} + h/3]$
Maximum vertical pressure q_{max}	$q_{max} = \rho\,S\,[z\,(z/A + 1)^{-1} + h/3]$	$q_{max} = \rho k\,S\,[z\,(z/A + 1)^{-1} + h/3]$	$q_{max} = \rho\,S\,[z\,(z/A_2 + 1)^{-1} + h/3]$
Lateral thrust p_z	$p_z = p_{max}\,[1 - ((z/A) + 1)^{-2}]$	$p_z = p_{max}\,[1 - ((z/A) + 1)^{-2}]$	Lateral thrust on the short side a: $p_z = p_{max}\,[1 - ((z/A_1) + 1)^{-2}]$ Lateral thrust on the long side b: $p_z = p_{max}\,[1 - ((z/A_2) + 1)^{-2}]$
Maximum lateral thrust p_{max}	$p_{max} = \rho\,D/(4tg\varphi')$	$p_{max} = \rho\,r/tg\varphi'$	$p_{max} = \rho\,a/(4tg\varphi')$ $p_{max} = \rho\,a'/(4tg\varphi')$ a: short side dimension b: long side dimension $a' = (2ab - a^2)/b$
Characteristic abscissa A	$A = D/(4\,tg\varphi'\,tg^2\,(\pi/4 - \varphi/2)) - h/3$	$A = L/\pi.1/[4\,tg\varphi'\,tg^2\,(\pi/4 - \varphi/2)] - h/3$	Characteristic abscissa for the small side a: $A_1 = 4a/\pi.1/[tg\varphi'\,tg^2\,(\pi/4 - \varphi/2)] - h/3$ Characteristic abscissa for the long side b: $A_2 = 4a'/\pi.1/[tg\varphi'\,tg^2\,(\pi/4 - \varphi/2)] - h/3$

Table 6.7. *Silo formulas*

The notations used in the above table have been defined as follows:

– ρ: volumetric weight of the silage material;

– φ: internal friction angle of the silage material;

– φ′: angle of friction of the material on the walls of the cell;

– S: inner surface of the cross-section of the cell;

– L: inner perimeter of the cell;

– D: inner diameter of the cell;

– r: average hydraulic radius of the section; r = S/L;

– h: height of the upper cone of silage material above the horizontal plane defined by z = 0;

– p: lateral thrust on the cell walls exerted by the ensiled material;

– q: vertical pressure exerted by the ensiled material on a horizontal slice;

– p_z: lateral thrust at depth z;

– q_z: vertical pressure at depth z;

– p_{max}: maximum lateral pressure;

– Q_{max}: maximum vertical pressure. $Q_{max} = S.q_{max}$;

– k: perimeter of the polygonal silo/perimeter of the circular silo.

6.10.3. *Overpressure in cells during emptying*

The tests carried out by Mr. Reimbert highlighted a significant increase in thrusts during the emptying of the ensiled material. The grains start moving with increasing speed from the outside toward the emptying axis. The experiment showed a collapsing cone of material with a funnel in the upper part.

This increase in pressure on the walls had already been mentioned by Mr. Prante during his tests: “The pressure on the walls is three to four times greater than that exerted by wheat at rest”.

From 1953 to 1954, full-scale tests were carried out on silos with direct measurements of the stresses on the walls using “resistant wire” gauges.

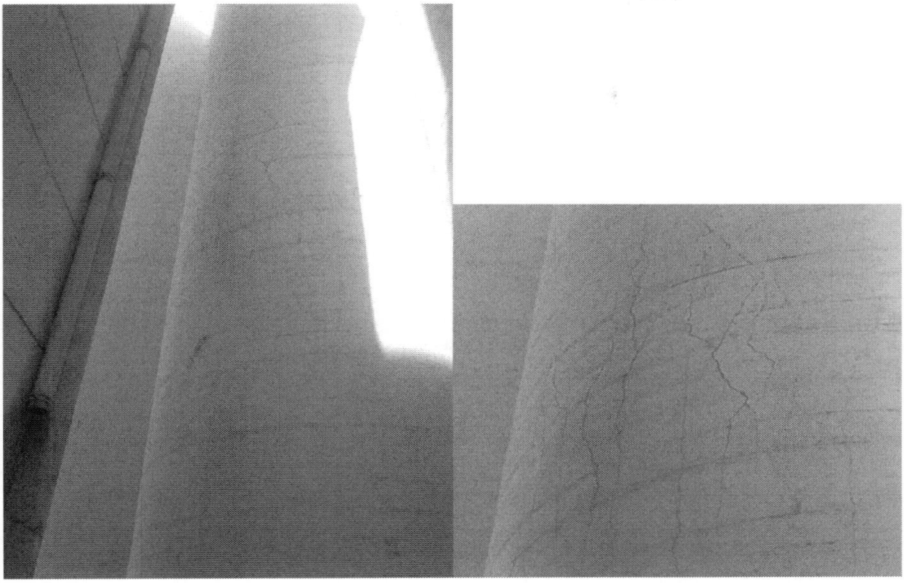

Figure 6.12. *Cell wall cracking due to increase in thrust when emptying. For a color version of this figure, see www.iste.co.uk/lauzin/concrete.zip*

These experiments showed that the overpressure coefficients were highly variable in the same silo and for the same ensiled material.

It also appeared that in silos presenting pathologies, the latter did not manifest themselves in the lower part of the cell but at mid-height.

The conclusion drawn from these tests by the commission is that calculating the structural dimensions could be carried out in two ways:

– by calculating the known efforts due to filling with the formulas above, increasing these efforts to consider emptying;

– by ignoring the increase in forces by equipping the cells with a "depression column", which allows for a laminar flow of the grains and a reduction in the pressure on the walls as emptying progresses.

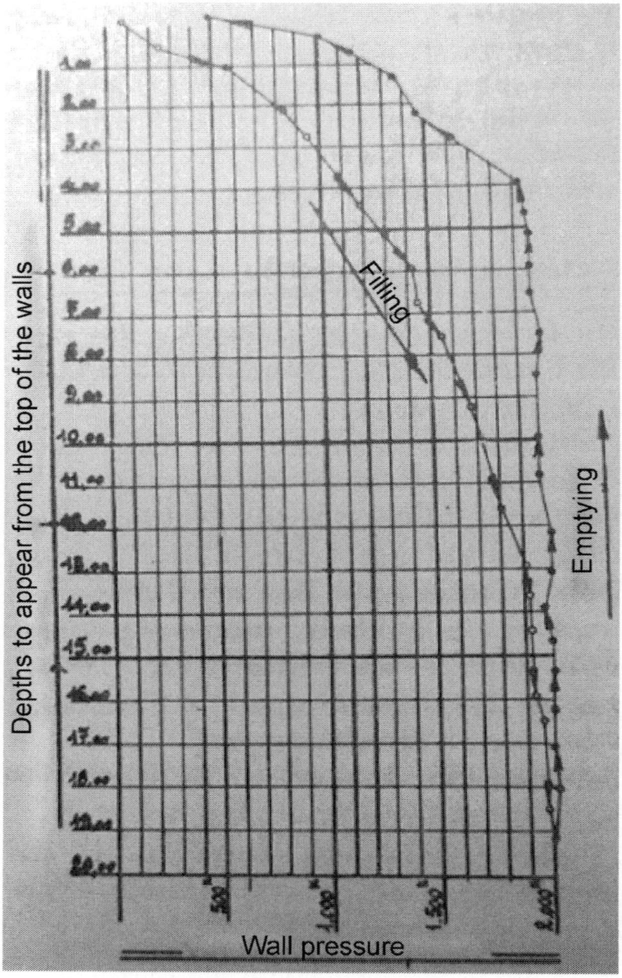

Figure 6.13. *Wall pressure values*

6.10.4. *Example of calculating the size of a square cell in reinforced concrete according to the previous formulas*

The calculation assumptions are as follows:

– rough walls;

– base dimension 2 m × 2 m;

– total height 18 m;

– silage material: wheat.

The values of the various parameters required for the calculation are deduced from Table 6.5:

– density of material between 750 and 840 kg/m^3;

– friction angle on rough walls $\varphi' = 25°$ (tg 25° = 0.466);

– minimum internal friction angle $\varphi = 24°$;

– maximum internal friction angle $\varphi = 26°$.

Calculation of characteristic abscissas is as follows:

– for the highest pressures: $A = L/\pi \times 1/[4 \, tg\varphi' \, tg^2 (\pi/4 - \varphi/2)] - h/3 = 4 \times 2/\pi \times 1/[4 \times 0.466 \times 0.307] = 4.45$;

– for the strongest lateral thrusts: $A = L/\pi \times 1/[4 \, tg\varphi' \, tg^2 (\pi/4 - \varphi/2)] - h/3 = 4 \times 2/\pi \times 1/[4 \times 0.466 \times 0.421] = 3.24$;

– hydraulic radius value: $r = S/L = 4/8 = 0.5$.

The maximum thrust is therefore $p_{max} = \rho \, r/tg\varphi' = 750 \times 0.5/0.466 = 805$ kg/m^2.

The maximum load on the bottom is as follows:

$$q_{max} = \rho \, kS \, [z \, (z/A + 1)^{-1} + h/3] = 840 \times 8/4\pi \times 4[17.534/(17.534/4.45 + 1) + 0.466 \times 2/(2 \times 3)] = 7{,}928 \text{ kg}$$

7

BA 60 Regulations (Modified in 1964)

7.1. Overview

As part of the BA 45 regulations, a periodic update was planned to oversee the technical developments of the materials.

The BA 60 regulations were based on the same concepts as those of 1945, with widespread use of high elastic limit steel.

The rules review commission was chaired by Mr. Caquot, with members renowned as M. Albigès, M. Lebelle, M. Esquillan, M. Perchat, etc.

The BA 60 regulations included eight chapters as follows:

– section 0: foreword;

– section 1: nature and quality of materials – tests;

– section 2: allowable stresses;

– section 3: general rules for strength calculations;

– section 4: special rules for certain elements;

– section 5: presentation of projects;

– section 6: execution of work;

– section 7: validation of work.

7.2. Calculation bases: requirements

7.2.1. *Basic requirements*

The following principles were considered:

– reinforced concrete structures must be designed to withstand the stresses for which they are intended even if cracks occur in the tension zones;

– cracking surfaces must be reinforced to ensure resistance and limitation of cracking;

– "the cracks are the consequence of the normal working of the structure" (EN 1992.1.1 section 7);

– reinforced concrete structures must be free from the possibility of brittle fracture.

NOTE.– The span of the beams was the length between the supports except for the beams on the bearing devices from where the central distance was taken.

7.2.2. *Project usage duration*

Project usage durations were not mentioned in the regulations.

However, section 0.1 "Regulation objectives" specified that they were designed to enable structures to be built with a degree of safety appropriate to their purpose and duration.

7.2.3. *Thermal effects*

Section 3.14 specified that the effects of shrinkage and temperature variations were not to be considered when the structure was intersected by expansion joints 25 m apart in dry regions or regions with strong temperature differences and 50 m apart in humid and temperate regions.

Beyond these values, thermal effects had to be considered in the calculations. The values of the shrinkage coefficients are given in the commentary to the text (Figure 7.1).

The thermal expansion coefficient was 10^{-5} m/m °C, and the temperature variation was ±20°C.

2.10^{-4} in Metropolitan France, north of the Bayonne-Grenoble line and in equatorial regions

3.10^{-4} in Metropolitan France, south of the Bayonne-Grenoble line and in North Africa in a 20km strip parallel to the coast

4.10^{-4} in North Africa beyond this strip

4.10^{-4} in the Sahara and Sudan

Figure 7.1. *Values of the shrinkage coefficients*

7.2.4. *Shrinking and differential movements*

There are no specific data on this subject in the Instruction.

7.3. Materials

7.3.1. *Concrete*

7.3.1.1. *Strength*

Section 1.2 of the regulations specified the following:

– the cement used belonged to class 250/315 unless justification was provided, proving that other categories could meet the requirements of the project;

– the recommended cement content was:

- 250 kg/m^3 for unreinforced or weakly reinforced concrete,

- 300 kg/m^3 for controlled standard concretes not exposed to bad weather or aggressive environments,

- 350 kg/m^3 for concretes of poorly controlled production or for structures not protected from the elements,

- 400 kg/m^3 for concrete intended for structures in aggressive atmospheres.

The regulations only applied to concrete with a compressive strength (measured on cylindrical specimens) greater than 15 MPa and a tensile strength greater than 1.5 MPa.

In the absence of prior studies, the admissible mechanical breaking strength at 28 days (in bar) is presented in Table 7.1.

Dosage	Uncontroled concrete		Controled concrete	
	σ_{28}	σ'_{28}	σ_{28}	σ'_{28}
250	150	15	180	18
300	190	17	230	20.5
350	225	18.5	270	22
400	250	20	300	24

Table 7.1. *Strength (bar) of concrete depending on dosage*

The compressive stress of the concrete $\bar{\sigma}_b$ was equal to a fraction of the compressive strength of the concrete at 28 days (σ_{28}), according to the following formula:

$$\bar{\sigma}_b = \alpha\beta\gamma\delta\varepsilon\ \sigma_{28}$$

with:

– α: it depends on the hardening rate of the concrete;

– β: it depends on the control exercised;

– γ: it depends on the ratio between the minimum thickness of the element (h_m) and the maximum size of the aggregates (c_g);

– δ: it depends on the stress distribution in the section.

Cement quality		**Cement 250/315**		**Cement 355/550**	
Value of coefficient α		1		5/6	
Concrete quality control		**Uncontrolled**		**Controlled**	
Value of coefficient β		1		6/5	
Element thickness		**$h_m \geq 4Cg$**		**$h_m \leq 4Cg$**	
Value of coefficient γ		1		$h_m/4C_g$	
Request	**Compression**	**Bending**	**Bending with traction**	**Bending with compression with $e_0 \geq 3i^2 / v'$**	**Bending with compression with $e_0 \leq 3i^2 / v'$**
Value of coefficient e δ	0.30	0.60	0.60	0.60	0.30 (1 + $e_0v'/3i^2$)
Value of coefficient ε	1	Between 1/2 and 1	Between 1/2 and 1	Between 1/2 and 1	Between 1/2 and 1

Table 7.2. *Coefficient values*

The values of the different coefficients are summarized in Table 7.2.

In typical cases, the strength values of the concrete samples are presented in Tables 7.3 (for uncontrolled concretes) and 7.4 (for controlled concretes).

Cement dosage	Allowable stress in simple compression (MPa)	Allowable stress in simple bending (MPa)	Permissible stress in simple traction (MPa)
250	4.5	9	0.45
300	5.7	11.4	0.51
350	6.75	13.5	0.55
400	7.5	15	0.60

Table 7.3. *Strength of uncontrolled concrete*

Cement dosage	Allowable stress in simple compression (MPa)	Allowable stress in simple bending (MPa)	Permissible stress in simple traction (MPa)
250	5.4	10.8	0.54
300	6.9	13.8	0.615
350	8.1	16.2	0.66
400	9	18	0.72

Table 7.4. *Strength of controlled concrete*

7.3.1.2. *Elastic deformation*

The longitudinal deformation modulus of the concrete was given by E_i and E_v as follows:

– $E_v = 7{,}000\ \sqrt{\sigma_j}$ for permanent loads and live loads;

– $E_i = 21{,}000\ \sqrt{\sigma_j}$ for rapidly varying loads (application duration less than 24 h).

If only 28-day tests were available on concrete, the values of E_v and E_i were those determined with σ_{28} increased by 20% for the 250/315 type concrete and by 10% for the other types.

Poisson's ratio was 0.15.

7.3.2. *Reinforced concrete steels*

7.3.2.1. *Properties*

The reinforced concrete steels considered in the regulations were as follows:

– smooth round bars (Adx and A42, medium-hard steel in A55, hard steel in A65 and A75);

– high/grip bars (natural or hardened steel);

– welded mesh;

– cut and stretched sheets.

With the exception of smooth round bars, which were standardized, the other products had to be the subject of an "approval form" guaranteeing their mechanical characteristics.

It was also permitted to use metal frame profiles.

The measured characteristics were as follows:

– the elastic limit σ'_e (apparent limit for natural steel and conventional limit at 0.2% for work-hardened steel);

– tensile strength σ'_{ar} (minimum 1.15 σ'_e);

– elongation at break ε'_{ar}.

7.3.2.2. *Strength*

– *Smooth round bars:* the common diameters were as follows: 5, 6, 8, 10, 12, 16, 20, 25, 32, and 40 mm. The mechanical characteristics are summarized in Table 7.5.

NOTE.– The elastic limit was 2,400 bar.

– *High-grip bars*: this was appropriate for referring to the approval sheets established by the manufacturer.

– *Welded mesh*: this mesh could be the subject of an approval form or meet the following characteristics:

- the main diameters of the mesh wires were 3, 3.5, 4, 4.5, 5, 6, 7, 8, 9, 10 and 12;

- the guaranteed mechanical characteristics are shown in Table 7.6.

	QUANTITATIVE CHARACTERISTICS			QUALITATIVE CHARACTERISTICS	
				COLD BENDING[s]	
	APPARENT ELASTIC LIMIT σ'_{eg} kgf/cm²	TENSILE STRENGTH σ'_{arg} kgf/cm²	ELONGATION AT BREAK ε'_{arg} %	CHUCK DIAMETER	BENDING ANGLE
A Dx	—	$3300 \leq \sigma'_{arg} \leq 5000$	—	4 ∅	90°
A 42	$\sigma'_{eg} \geq 2400$	$4200 \leq \sigma'_{arg} \leq 5000$	$\varepsilon'_{arg} \geq 25$	∅ ≤ 20: 0 ∅ > 20: ∅	180°
A 55	—	$5500 \leq \sigma'_{arg} \leq 6500$	$\varepsilon'_{arg} \geq 18$	—	—
A 65	—	$6500 \leq \sigma'_{arg} \leq 8500$	$\varepsilon'_{arg} \geq 12$	—	—
A 75	—	$\sigma'_{arg} \geq 7500$	$\varepsilon'_{arg} \geq 6$	—	—

Table 7.5. *Mechanical characteristics of smooth round steel bars*

Characteristics	Φ ≤ 6 mm	Φ ≥ 6 mm	Elongation at break
Minimum guaranteed value	6,300 bar	4,500 bar	Not guaranteed
Elastic limit	5,200 bar	4,250 bar	–

Table 7.6. *Mechanical characteristics of the welded mesh*

– *Cut and stretched sheets*: they had to be the subject of approval forms on which one had to find:

- geometric characteristics;
- sections to be taken into account in the calculations;
- tensile strength of the strips;
- the elastic limit to be taken into account in resistance calculations;
- constructive provisions for anchors and covers.

7.3.2.3. *Steel/concrete adhesion*

The steel/concrete adhesion breaking stress was as follows:

$$\tau_{rd} = \frac{2\eta_d^2 \sigma'_{28}}{\left(1+\frac{\eta_d \Phi}{d_1}\right)\left(1+\frac{\eta_d \Phi}{d_2}\right)}$$

with:

– Φ: nominal diameter (mm);

– d_1: minimum distance from the bar axis to the free surface of the concrete;

– d_2: smallest distance from the axis of the bar to the free surface of the concrete measured in both directions perpendicular to d_1 or as an extension of d_1;

– η_d: sealing coefficient of the bars ($\eta = 1$ for smooth round bars and $\eta = \sqrt{2}$ for high-grip bars);

– σ'_{28}: the permissible tensile stress of the concrete (bar) at 28 days.

The admissible adhesion stress was then:

$$\bar{\tau}_d = \frac{2\eta_d^2 \bar{\sigma}'_b}{\left(1+\frac{\eta_d \Phi}{d_1}\right)\left(1+\frac{\eta_d \Phi}{d_2}\right)}$$

where σ'_b is the permissible tensile stress of the concrete (bar).

The straight sealing lengths of the tensioned reinforcements were fixed in section 2.35 of the regulations by:

$$l'_d = \frac{\Phi}{4} \frac{\bar{\sigma}'_a}{\bar{\tau}_d}$$

where σ'_a is the allowable tensile stress of steel.

7.4. Durability and cover of reinforcements

7.4.1. *Overview*

The durability criteria are not mentioned in the Instruction, nor are the conditions of exposure of the structures to physicochemical attacks.

7.4.2. *Durability requirement*

Service lifetimes are not mentioned.

7.4.3. *Cover*

The covers were set to the following values:

– in a non-aggressive environment:

- 2 cm for facings exposed to bad weather, condensation or contact with liquids (tanks, pipes, conduits, etc.);

- 1 cm in other cases.

The mechanical cover was fixed to:

– $\Phi/2 + 8$ mm when the diameter Φ of the bar was less than 16 mm;

– Φ in other cases;

– in an aggressive environment, the cover was set in the CCTP. The regulations consider the following environments to be aggressive:

- pure water,

- waters loaded with calcium sulfate (selenitic water) or magnesium sulfate,

- seawater.

NOTE.– For marine environments, it was typical to consider a cover as equal to 35 mm.

7.5. Structural analysis

7.5.1. *Overview*

The purpose of structural analysis is to determine the distributions of stresses and deformations. The behavior model used in the Instruction is that of *linear elastic behavior*. The calculation methods were deterministic according to the principle of allowable stresses.

7.5.2. *Load combinations*

The calculations were carried out with:

– permanent loads (G_1 and G_2);

– shrinkage and temperature variations;

– static and dynamic operating live loads (P_1 and P_2).

The load combination applied was:

$$(G_1) + (G_2) + 1.2(P_1) + 1.2(P_2)$$

When the snow and the wind (P_3) were considered, the combination became:

$$(G_1) + (G_2) + 1.2(P_1) + 1.2(P_2) + (P_3)$$

However, in this case, the allowable stresses were multiplied by 9/8.

7.5.3. *Second-order effects*

Second-order effects that can significantly affect the *overall stability* of the structure were not considered in the Instruction.

7.5.4. *Structural modeling*

The digital modeling of structures was not provided by the regulations.

7.6. Structural calculations

7.6.1. *Compression simple*

The compression in the structure was calculated by considering the homogeneous section of the structure. The steel–concrete equivalence coefficient was then taken as 15. The elastic stability of the considered structure was also verified.

For the assessment of loads on column supporting continuous spans, an increase of 15% was applied for central columns (two spans), and an increase of 10% was applied for central columns adjacent to edge columns (more than two spans).

7.6.2. *Bending–compression with bending*

The tensile stresses in the steel were limited to:

$$\sigma'_a \leq \text{Min}\,(\sigma'_1; \sigma'_2)$$

where:

– σ'_1 corresponds to the mechanical resistance of steel;

– σ'_2 corresponds to the limitation of concrete cracking.

For structures subjected to bending, the value of this stress was given by:

$\sigma'_a \leq 2/3\ \sigma'_{ak}$ where σ'_{ak} is the elastic limit of the steel considered

In the general case, the limit imposed by the mechanical characteristics of steel and concrete was:

$$\sigma'_1 = \rho'_a\ \sigma'_{ak}$$

with:

– $\rho'_a = 3/5$ for longitudinal reinforcements subjected to simple traction (tie rods, hangers, tank or silo hoops), etc.;

– $\rho'_a = 3/5\ (1 + 1/9\ (e_0 \times v)/i^2)$ for longitudinal reinforcements subjected to composite bending (with tension). In this formula, e_0 is the eccentricity of the external force in the section, v is the distance from the center of gravity of the section to the less tense end and i is the radius of gyration;

– $\rho'_a = 2/3$ for longitudinal reinforcement of structures subjected to simple bending or composite bending with compression.

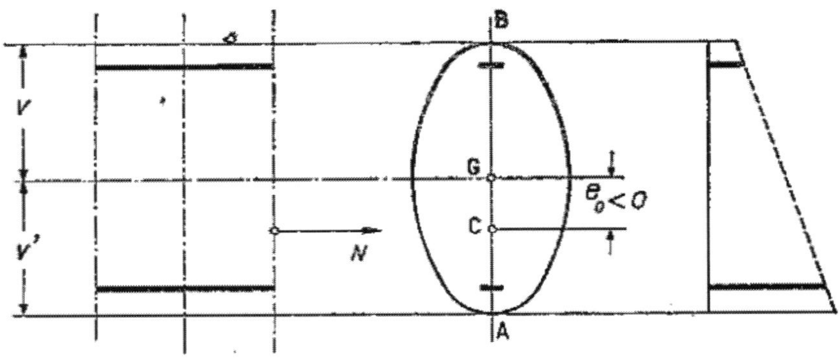

Figure 7.2. *Defining the section parameters*

In the case where the limit was imposed by the concrete cracking conditions, for example, for the tie rods and the reinforcement of the tank walls, the value of σ'_2 was given by the maximum of the following two values:

$$\bar{\sigma}_2' = \frac{k\eta\,\varpi_f'}{\Phi\left(10+\varpi_f'\right)}$$

$$\bar{\sigma}_2' = 7.5\sqrt{\frac{k\eta\,\bar{\sigma}_b'}{\Phi}}$$

with:

– Φ: the nominal diameter (mm) of the largest tensioned bar in the covered section (the covered section being defined as the area of the tensioned concrete);

– η: the cracking coefficient of the bars (η = 1 for smooth round bars and η = 1.6 for high grip bars);

– σ'_b: permissible tensile stress of the concrete (bar);

– k: a parameter that depends on the effects of cracking on the structural behavior (Table 7.7):

- cracking is not considered detrimental and does not compromise the preservation of the steel or the service life of the structure,

- cracking of stressed areas is considered detrimental when the structure is exposed to the elements, water, an aggressive atmosphere, or when the structure serves as a support for a fragile cover,

- the opening of cracks is considered very detrimental when the elements considered are in contact with water or exposed to a very aggressive environment or when they provide the sealing function (tanks, for example);

– $\omega'_f = 100\ A'/B'_f$, where A′ is the section of the tensioned bars and B'_f is the concrete cover section.

	Minimal cracking	**Detrimental cracking**	**Very detrimental cracking**
k	150,000	100,000	50,000

Table 7.7. *k values*

NOTE.– In the case where one face of the structure was not in permanent contact with water or an atmosphere saturated with water vapor, the values of the previous stresses were increased by 300η, that is:

$$\bar{\sigma}_2' = \frac{k\eta\,\varpi'_f}{\Phi\left(10+\varpi'_f\right)} + 300\eta$$

$$\bar{\sigma}_2' = 7.5\sqrt{\frac{k\eta\,\bar{\sigma}_b'}{\Phi}} + 300\eta$$

Generally, flexural elements are designed according to material resistance techniques. The simplified methods for floors with low live loads are detailed in section 4.221 of the regulations.

NOTE.– The calculations were also detailed for flexed walls (section 4.61) and wall-bearing beams (section 4.62).

7.6.3. *Tensile strength*

Under tensile stresses, the regulations recommended taking into account an "effective section", defined as the homogeneous section from which the tense zones of the concrete had been deducted.

The stresses obtained were compared with the allowable stresses.

7.6.4. *Buckling*

The method used to determine the buckling length l_c is described in section 4.15, which depends on the free length l_0 and is summarized in Table 7.8.

Assumption	**Free structure at one end and embedded at the other**	**Structure articulated at both ends**	**Structure embedded at both ends with possible horizontal movement**	**Structure embedded at one end and articulated at the other**	**Structure embedded at both ends without the possibility of movement**
l_c value	2 l_0	l_0	l_0	$l_0/\sqrt{2}$	$l_0/2$

Table 7.8. *Buckling length values*

NOTE.– In the case of multistory buildings, for the columns taken in continuity, the buckling length was 0.7 l_0 whether the column was embedded in the foundation block or assembled into a network of beams crossing it from side to side.

Slenderness value	≤ 50	≥ 50	
Verification criteria	Simple compression without taking buckling into account	Composite buckling under the effect of: – normal force N; – the bending moment defined as follows:	
ξ is the ratio of the permanent load to the total load; φ is the smallest of the distances from the center of gravity of the section of the structure reduced to concrete alone to the contour of this section in the plane where the buckling is; σ is the normal stress related to the homogeneous section under the effect of the normal force: $\sigma = \frac{N}{B + 15A}$ σ_E is the critical Euler stress related to the homogeneous section. $\sigma_E = \frac{\pi^2 E_b}{\lambda^2}$ where E_b is equal to: $E_\varphi + (E_i - E_\varphi) \times \frac{\text{rapidly varying live loads (}^1\text{)}}{\text{total load}}$		If $\sigma'_{AK} \geq 3\,600$ bar: the deformation of the mean line of the structure was: $f = \frac{1+\xi}{1300} \frac{(l_c - 50i)^2}{\varphi} \frac{\sigma_E}{\sigma_E - \sigma}$	If $\sigma'_{AK} \leq 3\,600$ bar: the deformation of the mean line of the structure was: $f = \frac{4(1+\xi)}{1\,600 + \sigma'_{ak}} \frac{(l_c - 50i)^2}{\varphi} \frac{\sigma_E}{\sigma_E - \sigma}$

Table 7.9. *Taking buckling into account*

In all other cases, this value was reduced to 0.9 l_0.

The slenderness of the piece had the same definition as in the previous regulations ($\lambda = l_c/i$).

The compressed structure considering buckling was verified, as indicated in Table 7.9.

7.6.5. *Punching shear*

The no-punching condition of a structure was defined by the following formula:

$$\tau_{max} = 1.5 \frac{Q}{p_c h_0}$$

where p_c is the perimeter of a contour parallel to that of the punching load but located at $h_0/2$ from the surface and h_0 is the thickness of the structure.

We will then have to check that:

$$\tau_{\max} = \leq \sigma'_b$$

NOTE.– Section 3.22 of the regulations is related to the implementation of a non-fragility percentage in the case of elements subjected to tensile stress.

Thus, for a bent beam, the percentage given for a basic section b is as follows:

$$\frac{100\,A'}{b_0 h} > \psi_4 \frac{\bar{\sigma}'_b}{\bar{\sigma}'_a} \left(\frac{h_t}{h}\right)^2$$

where $\bar{\sigma}'_b$ and $\bar{\sigma}'_a$ are the allowable tensile stresses of the concrete and steel, h_t is the total height, h is the useful height and ψ_d is a coefficient of value:

– 36 for soft and naturally hard steel;

– 54 for work-hardened steels.

For a bent joist, the values were as follows:

– for any p value

$$\frac{100\,A_x^{'}}{bh_x} \geq \frac{\varphi_4}{2}(2-p)\frac{\bar{\sigma}_b^{'}}{\bar{\sigma}_a^{'}}\left(\frac{h_0}{h_x}\right)^2$$

–for p ≤ 0.40

$$\frac{100\,A_y^{'}}{bh_y} \geq 0.35\varphi_4\frac{\bar{\sigma}_b^{'}}{\bar{\sigma}_a^{'}}\left(\frac{h_0}{h_y}\right)^2$$

–for $0.40 < p \leq 1$

$$\frac{100\,A_y^{'}}{bh_y} \geq \frac{\varphi_4}{4}(1+p)\frac{\bar{\sigma}_b^{'}}{\bar{\sigma}_a^{'}}\left(\frac{h_0}{h_y}\right)^2$$

where:

– $\rho = l_x/l_y$;

– A'_x and A'_y represent the sections of the reinforcements stretched by strips of width b following l_x and l_y;

– h_x and h_y are the useful heights of the joists.

7.6.6. *Torsion*

The verifications provided in the regulations are summarized in Table 7.10.

with:

– M_t: torque moment;

– a: diameter of the section or short side of the rectangular section;

– b: long side of the rectangular section;

– T: shear force;

– σ_{b0}: allowable stress of the concrete under simple compression;

– σ_b: normal compressive stress;

– ω_0: ratio of longitudinal torsion steels in the cross-section;

– σ'_a: = 3/5 σ'_{ak}.

Section type/stresses	Tangential stress	Verification	Longitudinal reinforcement	Transverse reinforcement
Circular section	$\tau_b = 5.09 \frac{M_t}{a^3}$	$\tau_b < \frac{\bar{\sigma}_{bo}}{3} - 1{,}5 \frac{T}{a^3}$ or $\sigma_b + \sqrt{\sigma_b^2 + 4\tau_b^2} < 2\bar{\sigma}_{bo}$	$\varpi_o = \frac{2}{3} \frac{\tau_b}{\bar{\sigma}'_a}$	$\varpi_{to} = 0.8 \frac{\tau_b}{\bar{\sigma}'_a}$
Rectangular section	$\tau_b = \frac{M_t}{a^2 b}\left(4.81 - 1.81 \frac{b-a}{\sqrt{2a^2 + b^2}}\right)$	$\tau_b < \frac{\bar{\sigma}_{bo}}{3} - \frac{4}{3} \frac{T}{ab}$ or $\sigma_b + \sqrt{\sigma_b^2 + 4\tau_b^2} < 2\bar{\sigma}_{bo}$	$\varpi_o = \frac{2}{3} \frac{a+b}{2b} \frac{\tau_b}{\bar{\sigma}'_a}$	$\varpi_{to} = 0.8 \frac{a+b}{2b} \frac{\tau_b}{\bar{\sigma}'_a}$

Table 7.10. *Torsion verification*

NOTE.– BA 60 regulations also presented calculation methodologies for:

– floors with moderate live loads;

– floors with high live loads;

– slabs on continuous supports;

– continuous beams, whether they are attached to the columns or not;

– bent walls.

NOTE.– The BA 60 regulations provided an approximate method for determining crack opening (as it would appear later in the Eurocodes and EN1992.1.1 in particular). The average value of crack opening when the latter was systematic was given by the expression:

$$w_m = \left(\sigma'_f - \frac{45\,\sigma'_j}{\varpi'_f}\right)\frac{3\Phi}{E_a\eta}\left(1 + \frac{10}{\varpi'_f}\right)$$

or even:

$$w = \frac{1}{500\,000\,\eta}\,\Phi\left[1 + \frac{10}{\varpi'_f}\right]\sigma'_f$$

where $\omega'_f = 100\ \sigma'_j/\sigma'_f$.

When the opening of the cracks was not systematic, the formula became:

$$w = \frac{\Phi\,(10 + \varpi'_f)\,\sigma'^2_f}{60\,\sigma'_j\,\eta\,E_a}$$

Steel stress (MPa)	Φ40 bar		Φ32 bar		Φ25 bar		Φ20 bar		Φ16 bar		Φ12 bar	
	BA 60	EC 2	BA 60	EC 2	BA 60	EC 2	BA 60	EC 2	BA 60	EC 2	BA 60	EC 2
160	0.3	0.4	0.2	0.3	0.2	0.2						
200			0.3	0.4	0.2	0.3			0.2	0.2		
240							0.3	0.4	0.2	0.3		0.2
280									0.3	0.4	0.2	0.3
320											0.3	0.4

Table 7.11. *Comparative table of crack opening (mm) between BA 60 and EC 2 for concrete with an average tensile strength of 2.9 MPa*

7.7. Constructive provisions

7.7.1. *Concrete pouring*

In the case of changes to straight sections, it was normal to make gussets with a slope of less than 1/3. Unless special arrangements were made, the concrete was not poured below 0°C.

7.7.2. *Distance between reinforcements and between reinforcements and formwork*

The minimum free horizontal distance between two bars of the same line was 6/5 C_g (where C_g is the maximum dimension of the largest aggregate). The vertical distance between a set of two bars and the third bar was as follows:

$$\text{Max}\ (3/4\ \Phi;\ 1/2\ C_g)$$

7.7.3. *Concrete joints*

Typically, concrete joints are constructed perpendicular to the direction of the compressive stresses. The second phase concrete then had to be poured and cleaned to "make the aggregates protrude". The joints had to be wetted.

7.7.4. *Concrete setting*

When necessary (thin, unprotected elements, etc.), the concrete was set by humidification or a temporary waterproof cover.

7.7.5. *Concrete control tests*

The control tests were carried out on cylindrical test pieces with a height equal to twice the diameter (generally, test pieces of 200 cm^2 with a cross-sectional height equal to 32 cm for a diameter of 16 cm). However, the use of cubic test pieces was permitted ($15 \times 15 \times 15$). These test pieces were kept for 24 h in a sealed mold in an enclosure at 20°C. After demolding, they were kept in water at a temperature of 20°C or in air at 95% humidity.

7.7.6. *Constructive arrangements of the reinforcements*

7.7.6.1. *Reinforcement overlap*

Overlapping of tensioned steel: In the same cross-section, the proportion of overlapping bars was:

– limited to half of the total design cross-section for structures subjected to bending;

– limited to one-third of the total design cross-section for structures subjected to simple tension or bending with tension.

The overlaps had to be joined by transverse reinforcements. The length of the straight overlap was given by:

$$1.2\, l_d^{'} \frac{\sigma_a^{'}}{\bar{\sigma}_a^{'}}$$

This value increases with the distance separating the two bars.

It was possible to place the curved anchor bars according to Figure 7.3.

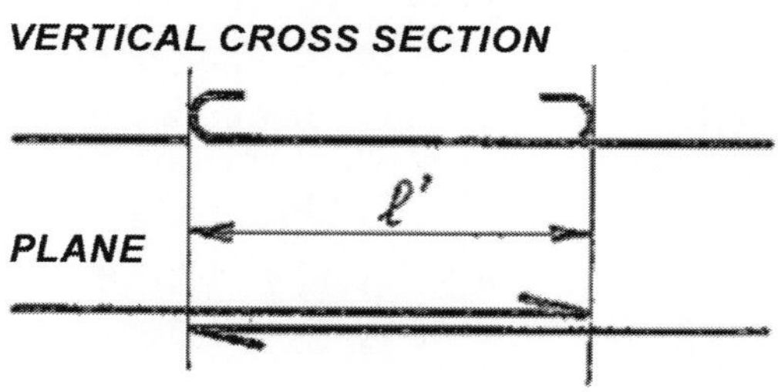

Figure 7.3. *Bar overlap with hooks*

In this case, the overlap length was set to:

$$0.6\, l_d^{'} \frac{\sigma_a^{'}}{\bar{\sigma}_a^{'}}$$

When the distance between the two bars was greater than 5Φ, the overlap length was increased by 5Φ.

For the welded mesh, the overlaps were three welds in the load-bearing direction and two in the distribution direction, with a minimum of:

$$2\,d_x + 10\Phi_x \qquad \text{if} \qquad \frac{\bar{\sigma}'_a}{3} < \sigma'_a \le \bar{\sigma}'_a$$

$$d_x + 10\Phi_x \qquad \text{if} \qquad 0 < \sigma'_a \le \frac{\bar{\sigma}'_a}{3}$$

where d_x is the distance between the axes of the distribution wires and Φ_x is the diameter of the carrier wires.

NOTE.– The continuity of steel by welding was permitted by the following regulations (section 2.38):

– the steels were weldable;

– the bars were welded end-to-end or by side beads with or without joints;

– welds were made using arc welding;

– the weld beads were sized according to CM rules (metal frame).

Threaded sleeve connections were also permitted.

7.7.6.2. *Columns*

The columns were reinforced with steel with an elastic limit greater than 360 MPa (smooth round bars or high-grip bars). The maximum distance between two neighboring bars on the same face was limited to 20Φ (Φ being the diameter of the smallest bar).

The minimum percentage of reinforcement of a column reduced to the volume of the column is:

$$\psi_1 \psi_2 \left(1 + \frac{2\,400}{\sigma'_{ak}}\right)$$

where σ'_{ak} is the elastic limit of the longitudinal reinforcements of the column and:

ψ_1 a coefficient taken equal to:

– 0.25 for the columns which receive at the upper level of the floor considered, either two beams following each other and serving themselves as support for two continuous floor spans, or four beams forming a cross;

– 0.35 for the columns that receive at the upper level of the floor considered, either two beams following each other but each of which supports only one floor span, or a single beam supporting two continuous floor spans, or three beams forming a T;

– 0.45 for the columns that support at the upper level of the floor considered, either a single beam loaded by a single floor span, or two orthogonal beams.

ψ_2 is a coefficient $(B/B_i)^2$ where B is the area of the section of the column and B_i is the area of the concrete section located inside the transverse reinforcements.

Column type	Square 30 × 30 or circular Φ30 column	Square 20 × 20 or circular Φ20 column	Square 14 × 14 or circular Φ14 column
ψ_2 value	1.77	2.44	3.84

Table 7.12. *ψ_2 values*

The minimum percentage of transverse reinforcement (in relation to the volume of the column) was:

$$0.15\left(1+\frac{2,400}{\sigma'_{ak}}\right)$$

In the frettage zones, the admissible compressive stress of the concrete can be increased by the following coefficient:

$$1+\mu_t\,\varpi_\varphi\left(1-\frac{2t}{a}\right)\frac{\sigma'_{ak}}{\sigma_{28}}$$

with:

– t: center distance of the transverse reinforcements;

– ω_φ: volume of transverse reinforcement per unit volume of concrete;

– a: smallest transverse dimension of the column, b largest dimension of the column;

– μ_t shape coefficient of the transverse reinforcements.

Frame shape	Helical frets	Square frames	Rectangular frames
Coefficient μ_t	3.00	1.50	1.50a/b

Table 7.13. *Values of the coefficient μ_t*

7.7.6.3. *Cover*

No less tolerance was allowed for the distances between the reinforcements and the covers.

7.8. Application to tank calculation

7.8.1. *Flat wall tanks*

Rectangular tanks could be sized using two methods:

– vertical slice method;

– horizontal slice method.

7.8.1.1. *Vertical slice method*

The principle consisted of cutting the tank into vertical sections delimited by two parallel planes spaced 1 m apart. The horizontal crosspiece (base or low slab) took the weight of the liquid, and the vertical walls behaved like two beams that took the thrust of the liquid. The short sides (transverse walls to the long sides) were calculated by considering horizontal sections. The vertical walls then underwent simple bending, whereas the horizontal wall was stressed via composite bending with traction.

When the tank rested on the ground, the weight of the liquid on the base was considered balanced by the reaction of the ground.

7.8.1.2. *Horizontal slice method*

The method used was that of the closed frame. The horizontal sections 1 m in height were then subjected to pressures in the form $p = \rho h$.

The moments at the ends of the walls were given by $M = -p/12\ (Ka^3 + b^3)/(Ka + b)$, with $K = I_b/I_a$ (the ratio of the moments of inertia of the long side and the short side), where a is the length of the short side and b is the length of the long side. Considering a section 1 m high, the values of the moments of inertia become $I_a = e_a^3/12$ and $I_b = e_b^3/12$, where e_a is thickness of side a and e_b is thickness of side b.

The mid-span moments were then given by:

$M_1 = pa^2/8 + M$ and $M_2 = pb^2/8 + M$

The shear force is then equal to T = pa/2 for wall AB. The walls were therefore subjected to a bending moment and a tensile force.

NOTE.– This method was preferred for slim tanks with small horizontal dimensions.

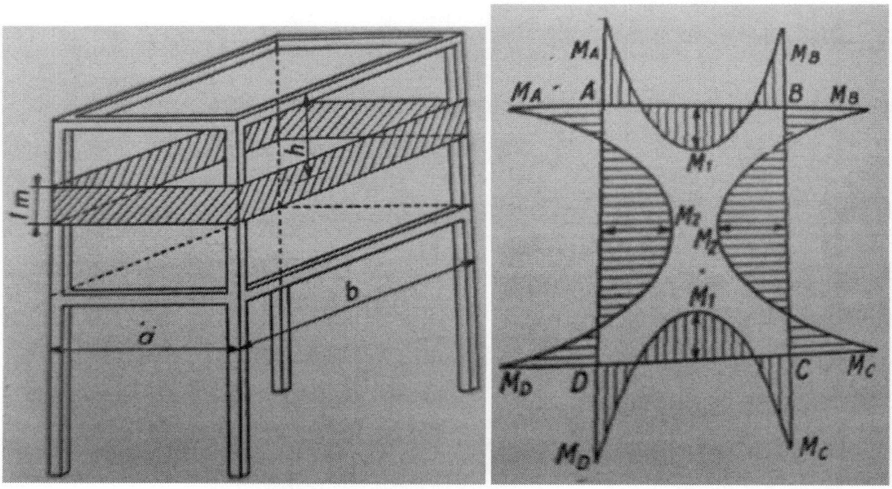

Figure 7.4. *Representation of bending moments*

7.8.1.3. *Design example*

We consider an 80 m^3 flat-walled tank resting on the ground. The dimensions are as follows: a = 4 m, b = 5 m and h = 4 m (height of the wall). The thickness of the walls varies from 30 cm at the base to 15 cm at the top of the wall. We consider the case of uncontrolled concrete dosed at 400 kg/m^3, that is, σ_b = 15 MPa and σ'_b = 0.6 MPa. With the steel being mild steel, cracking is considered very detrimental, which implies that K = 50,000 and n = 15.

The dimensions of the structure are calculated using horizontal slices. The calculation was often simplified by considering that the pressure exerted by the liquid on the walls was uniform and equal to the pressure calculated at the mid-height of the slice considered. The weighted pressure acting on the wall is:

$$p = 1.20\ \rho h$$

The stresses per 1 m height are then summarized in Table 7.14.

NOTE.– The moment at the ends M is given by: $M = -p/12\ (Ka^3 + b^3)/(Ka + b)$.

Slices (counted from the top of the tank)	Weighted pressure on the wall (daN/m²)	Moment at the ends of the slice (daNm)	Mid-range moment M_1 (daNm)	Mid-range moment M_2 (daNm)	Tensile strength on the short side (daN)	Tensile strength on the long side (daN)
0–1 m	600	–1,050	150	825	1,500	1,200
1–2 m	1,800	–3,150	450	2,475	4,500	3,600
2–3 m	3,000	–5,250	750	4,125	7,500	6,000
3–4 m	4,200	–7,350	1,050	5,775	10,500	8,400

Table 7.14. *Loads on the walls*

Since the thicknesses of the walls are equal on all four sides, we have K = 1. Therefore, with a = 4 and b = 5, M = –21 p/12 = –1.75 p.

The span moment is then deduced from the formula $M_1 = pa^2/8 + M$ and $M_2 = pb^2/8 + M$.

The concomitant traction is given by T = p × a/2 and T = p × b/2; thus, for the lowest slice, T = 4,200 × 4/2 = 8,400 daN.

The reinforcement was then determined by a composite bending calculation with M = –7,350 daNm and T = –8,400 daN.

7.8.2. *Circular tanks*

7.8.2.1. *Calculation method*

The case of circular tanks resting on a flat surface is considered here. These structures are generally covered by a circular flat slab or a dome. *Tank wall reinforcements* were calculated by dividing the latter into horizontal slices of 1 m in height on which a uniform pressure equal to the average pressure was applied.

The cross-section of the steel A_s necessary to balance the tensile force in the thickness of the wall was as follows:

$$A_s = pR/\sigma'_a$$

with:

– P: average pressure of the liquid in the considered slice;

– R: average radius of the tank;

– σ'_a: permissible stress of steel in tension.

Typically, the hoops were kept at a constant diameter along the section considered. To facilitate the building of the wall, it was also standard to keep the same diameter over the entire height of the wall and instead adjust the spacing of the bars.

The vertical reinforcement was generally composed of distribution reinforcements whose section was between 1/2 and 1/3 of that of the hoops. These bars were then spaced 10–12 cm at the foot of the wall and up to 20–24 cm above mid-height (one bar out of two stopped at mid-height).

The method used to calculate the hoops assumed that the deformation of all the slices was identical. As this assumption was not verified at the level of the wall/slab or wall/covering slab junction, some manufacturers took into account a fixed end moment of the wall on the slab, whose M_0 (kgm) value was given by:

$$M_0 = 3R \times H \times e$$

with:

– R: average radius (m) of the tank;

– H: total height (m) of the wall;

– e: wall thickness (cm).

This embedding at the base then made it possible to consider a horizontal axis bending moment M_1 (kgm), allowing the calculation of vertical reinforcements. The M_1 value was given by:

$$M_1 = M_0/5$$

The wall thickness was determined from the working rate of the tensioned concrete section, limiting the homogeneous section value to 20 bar (2 MPa). The formula used was as follows:

$$pR/(S + nA') \leq 20 \text{ bar}$$

with:

– S: concrete section of the wall;

– A′: section of the wall steel;

– n: steel–concrete equivalence coefficient.

As a general rule, it was empirically accepted that the section at the base of the wall was given at a minimum by the following formula:

$$e = H \times D/4$$

where H is the total height of the spire in m, D is the diameter of the tank in m and e is the thickness in cm. This value had to be greater than 8 cm.

7.8.2.2. *Design example*

Consider a circular tank 8 m in diameter and 5 m in height. The concrete has the characteristics indicated in Table 7.15 (for the uncontrolled concrete dosed at 400 kg/m^3).

Cement dosage	Allowable stress in simple compression (MPa)	Allowable stress in simple bending (MPa)	Allowable stress in simple traction (MPa)
250	4.5	9	0.45
300	5.7	11.4	0.51
350	6.75	13.5	0.55
400	7.5	15	0.60

Table 7.15. *Allowable stresses of concrete*

Slices (counted from the top of the dome)	h (m)	p (kg/m²)	T = pR (kg)	A′ (cm²) hoops	Actual cross-section (cm²) of the circle	Concrete traction (bars)
0–1 m	0.50	600	2,400	2	4Φ8	
1–2 m	1.50	1,800	7,200	6.10	8Φ10	
2–3 m	2.50	3,000	12,000	10.2	9Φ12	
3–4 m	3.50	4,200	16,800	14.3	13Φ12	
4–5 m	4.50	5,400	21,600	18.3	12Φ14	16.9

Table 7.16. *Steel section of the walls and concrete stress*

The reinforcement is composed of mild steel with an elastic limit equal to 2,400 bar. The structure was calculated to have very damaging cracking, that is,

K = 50,000, and the steel–concrete equivalence coefficient was taken as 15. The minimum thickness of the wall, given by the empirical formula, is 8 × 5/4 = 10 cm. A value varying linearly from *10 cm* at the base to *8 cm* at the top of the spire.

The tank wall is then broken down into five horizontal slices, each 1 m high. The weighted pressure of the liquid is equal to p = 1.20 × 1,000 × h = 1,200 h (assuming that the density of the liquid is equal to 1,000 kg/m^3). The calculation values are then summarized in Table 7.16.

NOTE.– The tensile stress of the steel was calculated for Φ =12 mm, that is, $\sigma'_a = \sigma'_2 = 1{,}180$ bar.

The distribution frames are composed of 12Φ8 per mL or 6.02 cm^2/mL. The platform was generally a 12-cm-thick slab reinforced with two 6Φ8 steel mesh reinforcements per milliliter.

7.9. 1964 Amendment (circular no. 70 of November 14, 1964)

The 1964 ministerial circular added only minor differences to the BA 60 regulations. These differences are mainly related to the following:

– *The mechanical resistance of the concrete to compression*: when only the 28-day test values were available, for values of j greater than 90 days, the admissible compressive stress values were as follows:

- $\sigma_j = 1.20\sigma_{28}$ for class 210–325 cement;
- $\sigma_j = 1.10\ \sigma_{28}$ for high-class cement.

– *The elastic limit of steel*: elastic limits were a function of the bar diameter according to Table 7.17.

– *Temperature effects*: for non-expandable structures, the value of the cover $\sigma = E \times \Delta l/l$ (relative expansion multiplied by the elastic modulus) was between:

- +20 and –20 bar in temperate climates;
- +30 and –30 bar in climates with high temperature variations.

– *Load combinations*: as in the BA 60 regulations, the live loads were increased by 20% unless the wind action was considered simultaneously. This means that the calculation combinations were as follows:

- first-order stress:

(G) + (T) + 1.2(P)

(G) + (T) + (P) + (N)

- second-order stress:

(G) + (T) + 1.5(P)

with:

– G: stress due to permanent load;

– T: stress due to thermal effects and shrinkage;

– P: stress due to live loads (including dynamic effects if necessary);

– V: stress due to wind or snow.

Steel grade	Steel diameter	Elastic limit
Mild steel	$\Phi \leq 20$ mm	2,400 bar
Mild steel	$\Phi = 25$ mm	2,300 bar
Mild steel	$\Phi \geq 25$ mm	2,200 bar
A42		2,400 bar
A55		3,200 bar
A65		3,600 bar
A75		4,000 bar

Table 7.17. *Elastic limit of steel*

The verification involved ensuring that:

– under first-order stresses, the stresses in the concrete and steel remain below the allowable stresses;

– under second-order stresses, the tensile stress of the steel remains below the elastic limit, and the compressive stress of the concrete is less than 1.5 times the allowable stresses;

– *allowable stresses*: the values of the coefficient δ are those indicated in Table 7.18.

In addition, the nominal strength value of the concrete must be greater than:

$$\sigma_{b0} \geq 90\sigma'_{ak}/(1{,}600 + \sigma'_{ak})$$

The values of the common elastic limits are listed in Table 7.19.

Stress	Compression	Bending	Bending with traction	Bending with compression $e_1 = 3i^2/(2v'-v)$
Value of coefficient δ	0.30	0.60	0.60	0.30 $(1 + e_0/e_1)$

Table 7.18. *Values of coefficient δ*

Elastic limit of steel (bar)	2,200	2,300	2,400	3,200	4,000	4,200	4,250	4,500	5,000
Concrete compressive stress (bar)	52.1	53.1	54	60	64.2	65.2	65.4	66.4	68.2

Table 7.19. *Values of the elastic limits of steel*

7.10. Tanks: 1966 CSNCCABP specifications

7.10.1. *Classification of structures – waterproofing criterion*

No waterproofing class is mentioned in the document. However, it is specified in section B, "Design of structures," that there are three possibilities:

– Type B1: reinforced concrete plays a mechanical role and a sealing role (cracking must be moderate according to the BA 60 regulations and section 2.123.1 of the recommendations);

– Type B2: reinforced concrete only plays a mechanical role, and sealing is ensured by an interior cover resistant to a certain amount of cracking;

– Type B3: reinforced concrete only plays a mechanical role but has a moderate opening of cracks, and sealing is ensured by an interior cover based on a hydraulic cover.

NOTE.– The recommendations were more specifically aimed at type B3 structures.

7.10.2. *Concrete composition and exposure class*

The 1966 specifications mainly included type B3 structures (cover applied to the internal side). The cement had to have:

– low shrinkage (slow setting cement of class 32.5 is recommended, and fast setting cements 42.5 and 52.5 are avoided). The use of cements with low hydration heat is recommended;

– sufficient mechanical and chemical resistance.

The minimum thickness of the concrete structures was set at 12 cm (except for small prefabricated elements, where it could reach 5 cm).

7.10.3. *Note on the justification of the tanks with BA 60*

For tanks with or without an adhesive interior cover (B1 and B3):

– the compressive stress of the concrete was limited to 30% of the compressive strength at 28 days;

– the tensile stress of the concrete was limited to:

$\sigma_b < \theta\ ft_{28}$ (concrete tensile limit at 28 days)

with:

– $\theta = 1$ (simple traction);

– $\theta = 1 + 2eo/3ho$ (composite bending);

– $\theta = 5/3$ (other cases);

– the tensile stress of the steels was limited to:

$$\text{Min}\ (3/5\ fe;\ 7.5\ (k\eta f_{tj}/\Phi)^{1/2} + 300\eta)$$

with:

– k = 30,000 for revolution tanks (admissible traction in the hoops);

– k = 50,000 for belt frames (angular points).

For tanks with a non-adherent interior cover (B2), the wall was not required to have any resistance to cracking. The structure was therefore designed to meet only mechanical resistance criteria.

As part of the recovery, it will therefore be necessary to plan for the implementation of non-adherent interior waterproofing (membrane type) or reinforced adhesive (rein-lined resin type) after checking the level of cracking.

Φ in mm	High grip steel with k = 30,000	High grip steel with k = 50,000	Smooth steel with k = 30,000	Smooth steel with k = 50,000	High grip steel from booklet 74 of 1998
6	220	268	165	204	220
8	197	239	147	181	197
10	181	219	135	165	181
12	170	204	126	154	170
16	153	183	113	137	153
20	142	168	104	126	142

Table 7.20. *Allowable tensile stresses in steel in MPa (category B1 and B3 structures) for 22 MPa concrete*

7.10.4. *Minimum constructive provisions*

The following provisions should be respected:

– the minimum diameter of the reinforcements was 6 mm for high-grip steel and 8 mm for smooth steel;

– the maximum distance between reinforcements was 20 cm;

– the reinforcements had to be arranged in two layers as soon as the thickness of the wall exceeded 15 cm (cylindrical tanks);

– gussets were provided at the sharp corners (15 cm × 15 cm minimum);

– the steel cover was set to:

- 3 cm for exterior facings and those in contact with water, condensation,
- 2 cm elsewhere.

Section 2.131.22 of the specifications specified the method to build this type of structure:

– installation of the exterior formwork to the height of one floor;

– reinforcement of walls and foundation;

– concreting of the foundation, lower gussets and wall;

– very quickly afterward, installation of the interior formwork leaving a free height at the bottom;

– large amount of water is sprayed into the formwork, which rapidly and continuously concretes over the entire height.

The free height for pouring fresh concrete was increased to 4 m. In comparison, this value is currently 1.50 m in booklet 65.

NOTE.– The 1966 specifications served as the basis for the development of booklet 74 of 1983.

7.11. Comparison between booklet 61 and the BA 60 regulations

The main differences are summarized in Table 7.21.

CPC – booklet 61	BA 60 regulations
Stresses	**Stresses**
Two types of stresses to consider. First order: (G) + (T) + 1.2(P) (G) + (T) + (P) + (V) Second order: (G) + (T) + 1.5(P) With: (G): stress due to permanent loads; (T): stresses due to temperature and shrinkage; (P): stress due to live loads including dynamic increase; (V): stress due to climatic loads. Under the first type of stresses, the stresses had to be less than the permissible stress limit. Under the second type of stresses, the stresses in the concrete had to be less than 1.5 × permissible stress limit and the stresses on the steels less than the elastic limit of the reinforcements.	1) For the verification of the active behavior, the stresses to be considered are: $(G_1) + (G_2) + 1.2(P_1) + 1.2(P_2)$ (1) $(G_1) + (G_2) + 1.2(P_1) + 1.2(P_2) + (P_3)$ (2) With: (G_1): stress due to permanent loads; (G_2): stresses due to temperature and shrinkage; (P_1): stress due to static operating loads; (P_2): stress due to loads involving a dynamic effect; (P_3): stress due to climatic effects. Under combination (1), the stresses shall not exceed the allowable stress limit. Under combination (2), the stresses shall not exceed 9/8 × allowable stress limit. 2) For safety verification, the stress is defined by: $(G_1) + (G_2) + \gamma_1(P_1) + \gamma_2 (P_2) + \gamma_3(P_3)$ (3) Under combination (3), the compressive stresses of the concrete should not exceed 1.5 × allowable stress and the tensile stresses on the steel less than the elastic limit of the reinforcements.

<table>
<tr><th>Allowable compressive stress of concrete</th><th>Allowable compressive stress of concrete</th></tr>
<tr><td>The permissible compressive stress of the concrete was:
$\bar{\sigma}_b = \alpha\beta\gamma\delta\varepsilon\ \sigma_{28}$
With:
$\alpha = 1$ for class 325 cement-based concrete;
$\alpha = 9/10$ for class 400 cement-based concretes;
$\alpha = 5/6$ for class 500 cement-based concretes;
$\beta = 1$ for strictly controlled concretes;
$\beta = 5/6$ for other concretes;
γ: the values are identical in both cases to the values given in section 7.3.1;
δ: the values are identical in both cases to the values given in section 7.3.1;
ε: in both cases, the value must be determined so that the average stress of the compressed zone, made homogeneous, does not exceed the admissible stress in simple compression (see section 7.3.1).</td><td>The permissible compressive stress of the concrete was:
$\bar{\sigma}_b = \alpha\beta\gamma\delta\varepsilon\ \sigma_{28}$
With:
$\alpha = 1$ for cement-based concrete of class 325;
$\alpha = 5/6$ for class 400 and 500 cement-based concretes;
$\beta = 6/5$ for strictly controlled concretes;
$\beta = 1$ for other concretes.</td></tr>
<tr><th>Reference tensile stress of concrete</th><th>Reference tensile stress of concrete</th></tr>
<tr><td>The reference tensile stress was given by:
$\bar{\sigma}'_b = \alpha\beta\gamma\theta\ \sigma'_{28}$
With θ coefficient linked to the cement dosage of the concrete:
<table>
<tr><th>Dosage</th><th>Θ</th></tr>
<tr><td>300</td><td>0.027</td></tr>
<tr><td>350</td><td>0.025</td></tr>
<tr><td>400</td><td>0.024</td></tr>
</table></td><td>The reference tensile stress was given by:
$\bar{\sigma}'_b = 0.30\ \alpha\beta\gamma\ \sigma'_{28}$</td></tr>
<tr><td>The resistance stresses attributable to concretes without prior testing are summarized in the table below (for a class 325 cement):
<table>
<tr><th>Cement dosage (kg/m^3)</th><th>σ_{28} (bars)</th><th>$\alpha\sigma'_{28}$ (bars)</th></tr>
<tr><td>300</td><td>230</td><td>20.5</td></tr>
<tr><td>350</td><td>270</td><td>22</td></tr>
<tr><td>400</td><td>300</td><td>24</td></tr>
</table>
Note: If the concrete is not subject to strict control, these values should be multiplied by 5/6.</td><td>The resistance stresses attributable to concretes without prior testing are summarized in the table below (for a class 325 cement):
<table>
<tr><th>Cement dosage (kg/m^3)</th><th>σ_{28} (bars)</th><th>$\alpha\sigma'_{28}$ (bars)</th></tr>
<tr><td>250</td><td>180</td><td>18</td></tr>
<tr><td>300</td><td>230</td><td>20.5</td></tr>
<tr><td>350</td><td>270</td><td>22</td></tr>
<tr><td>400</td><td>300</td><td>24</td></tr>
</table>
Note: If the concrete is not subject to strict control, these values should be multiplied by 5/6.</td></tr>
</table>

<table>
<tr><th>Allowable stress of steels</th><th>Allowable stress of steels</th></tr>
<tr>
<td>The permissible stresses of the steels were determined from the nominal yield loads in compression and tension defined as the mean values of the real yield loads reduced by twice the root mean square deviation.
The nominal yield loads σ'_{en} commonly accepted were the following:
<table>
<tr><th>Steel grade</th><th>Nominal yield load (bars)</th></tr>
<tr><td>Adx $\phi \leq 20$ mm</td><td>2,400</td></tr>
<tr><td>Adx $\phi = 25$ mm</td><td>2,300</td></tr>
<tr><td>Adx $\phi > 25$ mm</td><td>2,200</td></tr>
<tr><td>A42</td><td>2,400</td></tr>
<tr><td>A55</td><td>3,200</td></tr>
<tr><td>A65</td><td>3,600</td></tr>
<tr><td>A75</td><td>4,000</td></tr>
</table>
Note: For high grip bars, the values were defined in the approval sheets.</td>
<td>The permissible limit stresses of steels are determined from the yield loads.
The yield loads σ'_{ak} commonly accepted were the following::
<table>
<tr><th>Steel grade</th><th>Yield load (bars)</th></tr>
<tr><td>Adx</td><td>2,400</td></tr>
<tr><td>A42</td><td>2,400</td></tr>
<tr><td>A55</td><td>3,200</td></tr>
<tr><td>A65</td><td>3,600</td></tr>
<tr><td>A75</td><td>4,000</td></tr>
</table>
Note: For high-grip bars, the values were defined in the approval sheets.
For welded meshes with a diameter of the supporting wires less than 6 mm, this value was 5,200 bar. For other wire diameters, it was 4,250 bar.</td>
</tr>
<tr>
<td>The permissible tensile stress $\overline{\sigma}'_a$ was given by the formula:
$\overline{\sigma}'_a = \rho'_a \sigma'_{en}$
With: $\rho'_a = 2/3$ except for transverse reinforcements and cases where cracking is detrimental.</td>
<td>The permissible tensile stress $\overline{\sigma}'_a$ was given by the formula:
$\overline{\sigma}'_a = \rho'_a \sigma'_{ak}$
The values of ρ'_a are defined in section 7.6.</td>
</tr>
<tr>
<td>The admissible adhesion stress $\overline{\tau}_b$ reinforcements with respect to the concrete was given by:
<table>
<tr><th>Steel grade</th><th>Situation</th><th>$\overline{\tau}_b$ Value</th></tr>
<tr><td>Smooth round</td><td>Beams</td><td>$2\,\overline{\sigma}'_b$</td></tr>
<tr><td></td><td>Plates and shells</td><td>$2.5\,\overline{\sigma}'_b$</td></tr>
<tr><td></td><td>Anchoring zone</td><td>$1.3\,\overline{\sigma}'_b$</td></tr>
<tr><td>High grip bars</td><td>Beams</td><td>$3\,\overline{\sigma}'_b$</td></tr>
<tr><td></td><td>Plates and shells</td><td>$3.75\,\overline{\sigma}'_b$</td></tr>
<tr><td></td><td>Anchoring zone</td><td>$2.8\,\overline{\sigma}'_b$</td></tr>
</table></td>
<td>The admissible adhesion stress $\overline{\tau}_b$ reinforcements with respect to the concrete was given by:
<table>
<tr><th>Steel grade</th><th>Situation</th><th>$\overline{\tau}_b$ Value</th></tr>
<tr><td>Smooth round</td><td>Beams</td><td>$2\,\overline{\sigma}'_b$</td></tr>
<tr><td></td><td>Plates and shells</td><td>$2.5\,\overline{\sigma}'_b$</td></tr>
<tr><td></td><td>Anchoring zone</td><td>$1.1\,\overline{\sigma}'_b$</td></tr>
<tr><td>High grip bars</td><td>Beams</td><td>$3\,\overline{\sigma}'_b$</td></tr>
<tr><td></td><td>Plates and shells</td><td>$3.75\,\overline{\sigma}'_b$</td></tr>
<tr><td></td><td>Anchoring zone</td><td>$2\,\overline{\sigma}'_b$</td></tr>
</table></td>
</tr>
</table>

Table 7.21. *Comparison between booklet 61 of CCTG and the BA 60 regulations*

8

CCBA 68

8.1. Overview

These regulations were validated by Decree No. 68–340 of April 4, 1968. They were applicable to all reinforced concrete constructions. They were the subject of booklet 61 title VI of the common specifications (CPC) applicable to public structures or building contracts awarded in the name of the State and replaced booklet 61 title VI published in 1964 and rules BA 60.

They were published in the form of a unified technical document (UTD). They remained in force with a few minor updates until the early 1980s with the publication of BAEL.

Compared with the 1964 version, the main changes were as follows:

– the addition of climatic, seismic, live, and service load effects (chapter I);

– the mechanical characteristics of the drawn wire, welded mesh, and other types of steel (chapter II);

– anchoring welded mesh (chapter VI).

They were structured around the following chapters:

– Chapter I: Evaluation of load requests;

– Chapter II: Mechanical characteristics of materials and allowable stresses;

– Chapter III: Principles of justification and general rules;

– Chapter IV: Normal stresses on straight sections of prismatic structures;

– Chapter V: Tangent or punching actions;

– Chapter VI: Adhesion;

– Chapter VII: Columns;

– Chapter VIII: Simply bent beams;

– Chapter IX: Panels on continuous supports;

– Chapter X: Localized pressures, hooping and articulations;

– Chapter XI: Various construction provisions;

– Chapter XII: Cracking;

– Chapter XIII: Deformations;

– Chapter XIV: Provisions applicable to current buildings;

– Chapter XV: Miscellaneous provisions;

– Chapter XVI: Notations;

– Appendices.

These regulations were only applicable to standard reinforced concrete structures; they did not apply to structures made of lightweight reinforced concrete or reinforced cavernous concrete.

They represent a transition between regulations with allowable stresses and semi-probabilistic regulations (BAEL).

8.2. Calculation bases: requirements

8.2.1. *Basic requirements*

Three types of checks were required in the regulations:

– verification of static balance (under first- and second-order stresses);

– verification of the mechanical resistance of all the structures (parts and assemblies);

– verification of the shape stability of slender elements.

It was also envisaged, where necessary, to justify:

– concrete cracking;

– deformations.

8.2.2. *Project duration*

The regulations did not define a project service life. The cover defined in Article 43 (4 cm in a marine atmosphere and 2 cm for facings exposed to the elements), however, suggested that the service life was less than 50 years (by applying the formula linking the carbonation depth to the square root of time, the service life would be approximately 30 years).

8.2.3. *Thermal effects*

Article 3 referred to the NV 65 rules (as amended in 67).

The comments to Article 4 specified the following values:

– For non-massive constructions, thermal effects were taken into account in the form $\lambda \times \Delta l/l$ (product of the thermal expansion coefficient of the concrete by the relative elongation of the structure), the value of which was between 20 bar and –20 bar for temperate climates and between 30 bar and –30 bar for climates with a large thermal difference.

– In other cases, a thermal expansion coefficient of concrete was 10^{-5} m/m/°C.

The thermal effects (shrinkage and expansion) of freely expandable structures were not considered.

Article 51 then sets the maximum lengths of the current buildings at:

– 25 m for dry regions or regions with strong thermal gradients;

– 50 m for humid or temperate regions.

8.2.4. *Settlement and differential movements*

The regulations did not provide information on this topic.

8.2.5. *Material properties*

Concrete shrinkage: The effects of shrinkage were not taken into account in freely expandable structures. In other cases, the tensile stress resulting from free shrinkage was as follows:

– 20 bar in the northern half and southwestern quarter of France and in equatorial climates;

– 30 bar in the southeastern quarter of France less than 20 km from the Mediterranean coast and in a humid tropical climate;

– 40 bar more than 20 km from the southern and eastern coasts of the Mediterranean and in a hot and dry climate;

– 50 bar in a very dry tropical or desert climate.

NOTE.– For non-massive open-air constructions, the shortening values due to shrinkage were:

– 2×10^{-4} m/m in the northern half and southwestern quarter of France and in an equatorial climate;

– 3×10^{-4} m/m in the southeastern quarter of France less than 20 km from the Mediterranean coast and in a humid tropical climate;

– 4×10^{-4} m/m greater than 20 km from the southern and eastern coasts of the Mediterranean and in a hot and dry climate;

– 5×10^{-4} m/m in a very dry tropical or desert climate.

– concrete creep: the concept of creep was not addressed in the regulations.

8.3. Materials

8.3.1. *Concrete*

The concrete strength value to be considered in the calculations is known as the "nominal strength". This nominal value was estimated from the value resulting from the arithmetic mean being reduced by 8/10 × standard deviation. For example, $\sigma'_m = \sigma'_b = \Sigma\sigma'/n - 0.80\ \Delta\sigma'_n$.

8.3.1.1. *Strength*

Compressive strength

The strength of the concrete was characterized by its simple compressive breaking stress measured on cylindrical specimens with cross-sections of 200 cm² and a slenderness of 2. The value was measured at 28 days. The allowable compressive stress of the concrete $\bar{\sigma}'_b$ was equal to a fraction of the nominal strength. That is:

$$\bar{\sigma}'_b = \rho'_b\, \sigma'_b$$

where:

$$\rho'_b = \alpha\beta\gamma\delta\varepsilon$$

with:

– α: it depends on the hardening rate of the concrete;

– β: it depends on the control exercised;

– γ: it depends on the ratio between the minimum thickness of the element (h_m) and the maximum size of the aggregates (c_g);

– δ: it depends on the stress distribution in the section.

The values are summarized in Tables 8.1–8.4.

Cement quality	Cement class 325	Cement class 400	Cement class 500
Value of coefficient α	1	9/10	5/6

Table 8.1. *Values of coefficient α*

Concrete quality control	Attenuated control	Strict control
Value of coefficient β	5/6	1

Table 8.2. *Values of coefficient β*

Element thickness	$h_m \geq 4C_g$	$h_m \leq 4C_g$
Value of coefficient γ	1	$h_m/4C_g$

Table 8.3. *Values of coefficient γ*

Stress	Compression	Bending	Bending with traction	Bending with compression with $e_0 \geq 3i^2/v'$	Bending with compression with $e_0 \leq 3i^2/v'$
Value of coefficient δ	0.30	0.60	0.60	0.60	$0.30\,(1 + e_0v'/3i^2)$
Value of coefficient ε	1	Between 1/2 and 1	Between 1/2 and 1	Between 1/2 and 1	Between 1/2 and 1

Table 8.4. *Values of coefficients δ and ε*

Tensile strength

The reference tensile stress $\bar{\sigma}_b$ was given by the fraction of the nominal stress according to the following formula:

$$\bar{\sigma}_b = \rho_b \, \sigma'_b$$

where:

$$\rho_b = \alpha\beta\gamma\theta$$

where α, β and γ maintain the same values as those for the compressive strength. θ takes the following values:

– $0.018 + 2.1/\sigma'_b$ for a first-order weighted load;

– $1.5\,(0.018 + 2.1/\sigma'_b)$ for a second-order weighted load.

In common cases, the strength values of concretes were as follows:

– for concretes with attenuated control: see Table 8.5;

Cement dosage	Allowable stress in simple compression (MPa)	Allowable stress in simple bending or bending with tension (MPa)	Allowable stress in simple traction (MPa)
250	4.5	9	0.45
300	5.7	11.4	0.51
350	6.75	13.5	0.58
400	7.5	15	0.62

Table 8.5. *Resistance of concrete in attenuated control*

– for concretes with strict control: see Table 8.6.

Cement dosage	Allowable stress in simple compression (MPa)	Allowable stress in simple bending or bending with tension (MPa)	Allowable stress in simple traction (MPa)
250	5.4	10.8	0.54
300	6.9	13.8	0.62
350	8.1	16.2	0.70
400	9	18	0.75

Table 8.6. *Concrete strength under strict control*

8.3.1.2. *Elastic deformation*

The rules defined an instantaneous longitudinal deformation modulus E_i from the compressive strength of the concrete at j days expressed in bar using the following formula:

$$E_i = 21{,}000 \sqrt{\sigma'_j} \text{ (bars)}$$

Similarly, there was a deferred module whose value was given by:

$$E_j = 7{,}000 \sqrt{\sigma'_j} \text{ (bars)}$$

For large values of j, it was assumed that:

– $\sigma'_j = 1.20\ \sigma'_m$ for concrete made from the 325 cement class;

– $\sigma'_j = 1.10\ \sigma'_m$ for concretes made from higher class cements.

8.3.2. *Reinforced concrete steels*

8.3.2.1. *Properties*

The 1968 regulations defined the mechanical characteristics of steels based on their elastic limits. The main steels used were:

– smooth steels of grades FeE22, FeE24 and FeE34;

– high-grip steels of grades FeE40 (A and B), FeE45 and FeE50;

– drawn wires;

– welded mesh.

The values of the elastic limits are summarized in Table 8.7.

Class	Elastic limit σ_{en} (bar)
FeE22	2,200
FeE24	2,400
FeE34	3,400
FeE40 (diameters from 5 to 16, 20)	4,200
FeE40 (diameters from 25 to 32, 40)	4,000
FeE45 (all diameters)	4,500
FeE50 (diameters from 5 to 16)	5,000

Table 8.7. *Elastic limits of common steels*

NOTE.– The nominal yield strength corresponded to an elongation of 0.2% of the bar. The elasticity coefficient of steel Ea was 2,000,000 bar.

The values used for the welded mesh are listed in Table 8.8.

Nominal diameter	Elastic limit (bar)	Tensile strength (bar)
3	5,200	6,000
3.5	5,200	6,000
4	5,200	6,000
4.5	5,200	6,000
5	5,200	6,000
6	5,200	6,000
7	4,410	5,100
8	4,410	5,100
9	4,410	5,100
10	4,410	5,100
12	4,410	5,100

Table 8.8. *Elastic limits and tensile strength of common welded meshes*

8.3.2.2. *Strength*

The allowable stress of steel under bending or compression $\bar{\sigma}_a$ was taken as equal to a fraction of the elastic limit σ_{en} according to the following formula:

$$\bar{\sigma}_a = \rho_a \sigma_{en}$$

For the values of ρ_a, see Table 8.9.

Stresses	First order	Second order
Value of $\boldsymbol{\rho_a}$	2/3	1

Table 8.9. *Permissible stress of steel according to the stresses*

NOTE.– The use of raw drawn wires was permitted only in the form of factory prefabricated elements assembled by welding. The use of high-grip drawn wires was permitted only under the above conditions. The permissible tensile stresses of these drawn wires were defined in the steel identification sheets. The use of these wires was not permitted in compression.

8.3.2.3. *Steel/concrete adhesion*

The bond stresses τ_d for the bar drive are given by:

$$\tau_d = T/z.p_{ui} \times A_i/A$$

with:

– A: total section of the tensioned reinforcements;

– A_i: section of the bar studied i;

– p_{ui}: useful perimeter of bar i (for a single bar, it is equal to the nominal perimeter, for a set of two bars to the circumscribed perimeter, and for a set of three bars to $2\pi\Phi/3$);

– T: punching force value;

– z = 7/8 h for rectangular sections and T sections.

The verification of the bond stresses was carried out under first-order load effects.

The allowable bond stresses were as follows:

– $2\psi_d\ \bar{\sigma}_b$ for the beams;

– $2.5\ \psi_d\ \bar{\sigma}_b$ for walls and panels.

The value of ψ_d was set to 1 for smooth round bars and to the value on the identification cards for high-grip bars.

The straight sealing length l_d of a bar Φ in diameter necessary for the transmission of all the forces to the concrete was then:

$$l_d = \Phi/4 \times \sigma_{a\ adm}/\tau_d$$

For a bar subjected to tensile stress σ_a, the formula for the straight sealing length l became:

$$l = \Phi/4 \times \sigma_a/\tau_d$$

For the welded meshes, each transverse wire weld could balance at most one-third of the allowable forces on the carrier wire and half on the distribution wire. The straight seal was ensured by at least three transverse steel welds for the carrier wires and two welds for the distribution wires. The mesh covering had to include three welds on each overlapping wire and two distribution welds.

8.4. Durability and cover of reinforcements

8.4.1. *Overview*

The sustainability criteria are not mentioned in the Instruction, nor are the conditions of exposure of the structures to physicochemical attacks.

8.4.2. *Durability requirement*

The rules did not provide recommendations on the service life of the structures.

8.4.3. *Cover*

In addition to the mechanical cover described in section 9.7.2, the protection of steels described in article 43 of the regulations was set at:

– 4 cm for work at sea or exposed to salt spray;

– 2 cm for facings exposed to bad weather, condensation, or in contact with liquids;

– 1 cm for walls located in covered rooms and not exposed to condensation.

8.5. Structural analysis

8.5.1. *Overview*

The purpose of structural analysis is to determine the distribution of load effects or stresses and deformations. The behavior model used in the Instruction is that of *linear elastic behavior*.

For BA 60, the calculation methods were predetermined according to the principle of allowable stresses. However, when the special requirements requested, Chapter XV allowed the sections to be justified by a limit state calculation (failure calculation). In this case, reference was then made to the practical recommendations of the European Concrete Committee (the CEB was created in 1953 by the National Trade Union Chamber of Reinforced Cement and Prestressed Concrete Constructors which tried to unify the calculation methods in Europe). In these recommendations, the semiprobabilistic theory classified limit states into two categories:

– the ultimate limit state that allowed the maximum value of the load-bearing capacity to be verified;

– the limit use states based on normal operating conditions or *durability*.

The method adopted could be defined as a semiprobabilistic limit state method resulting from a combination of deterministic and probabilistic methods. The process is as follows:

1) Taking into account the resistance of materials and actions in *characteristic values*, each value is determined according to the probability that the actual values are lower or higher than the chosen values.

2) Taking into account uncertainties by transforming these characteristic values into *calculation values* in the following way:

- for materials, the reduction coefficients of the mechanical properties are introduced;

- for actions and stresses, by introducing increase or decrease coefficients according to the probability of occurrence.

3) Correcting the calculation values by introducing behavior coefficients to consider the behavior of the structure or the nature of the materials (brittle fracture, etc.).

The recommendations, also dealing with the prestressed concrete component, proposed two options with regard to cracking:

– *Do not recognize cracking*: the structures were then divided into two classes (class I or class II). Class I mainly targeted structures subjected to dynamic actions, fatigue and those particularly exposed to corrosion or requiring sealing (tanks, etc.). This class mainly targeted prestressed concrete. Class II, for which the decompression limit state of the concrete could only be verified under a fraction of the applied loads, was intended for prestressed concrete structures but also for reinforced concrete structures to the extent that the tensile stresses were sufficiently low to satisfy the required criteria.

– *Recognize cracking as a normal phenomenon*: the structures were also divided into two classes (class III and class IV). Class III corresponded to partial prestressing, and class IV corresponded to conventional reinforced concrete structures. The values retained for the crack opening were w = 0.1, 0.2, or 0.3 mm, depending on the type of structure and the aggressiveness of the environment.

8.5.2. *Load combinations*

CCBA 68 requires that the justification of sections be carried out under the following load types:

– the total first-order balanced load effects, the formulations of which are as follows:

- $(S_1) = (G) + 1.2(P) + (T)$;

- $(S'_1) = (G) + (P) + (V) + (T)$.

These loads were mainly used to calculate reinforcements;

– the total second-order balanced load effects, the formulations of which are as follows:

- $(S_2) = (G) + 1.5(P) + (T) + 1.5(V)$;

- $(S'_2) = (G) + (P) + \gamma_w(W) + (T)$;

- $(S''_2) = (G) + (P) + (SI) + (T)$;

– the service load effects were calculated as part of the justification relating to concrete cracking and structural deformations. They are defined by the following formula:

$$(S) = (G) + (P) + (V) + (T)$$

In these formulas, (G) represents the stress due to the permanent load, (P) represents the stress due to operating loads, including any increases in dynamic effects, (V) represents the stress due to normal climatic live loads, (W) represents the stress due to extreme climatic live loads, (T) represents the stress due to the effects of temperature and shrinkage, and (SI) represents the stress due to earthquakes.

The coefficient γ_w has the following value:

– $1.10 - 0.5 \times (P_{g.max}/(G))$ if $P_{g.max} \leq 0.20(G)$;

– 1 otherwise.

$P_{g.max}$ represents the maximum load effect due to operating live loads.

The first type of stress considered was the most common for current structures; only the operating loads were balanced to take into account possible variations. The second type of stress concerned the verification of the safety of the structure with respect to live loads.

NOTE.– CCBA 68 also considered the case of temporary constructions, for which the total balanced load effects were then as follows:

– first-order load effects: $(G) + (P) + (V) + (T)$;

– second-order load effects: $(G) + 1.25\ (P) + (V) + (T)$ and $1.3(G) + (T)$.

8.5.3. *Second-order effects*

Second-order effects that can significantly affect the *overall stability* of the structure were not considered in the regulations.

8.5.4. *Structural modeling*

CCBA 68 does not target the digital modeling of structures.

8.6. Calculation of structures

8.6.1. *Simple compression*

The regulations define a minimum section of longitudinal reinforcement equal to:

$$A'_m = 1.25/1{,}000\ \theta_1\theta_2\theta_3\ N'/\ \bar{\sigma}'_{b0}$$

and transverse reinforcements equal to:

$$A'_t = 1.5/1{,}000\ \theta_1\theta_2\ N'/\bar{\sigma}'_{b0}$$

with:

– **N′**: normal force applied to the structure;

– σ'_{b0}: allowable stress of the concrete in compression;

– θ_1: values shown in Table 8.10;

– $\theta_2 = 1 + l_c/(4a - 2c)$, with l_c: buckling length; a: small dimension of the cross-section; and c: cover of the longitudinal reinforcements;

– $\theta_3 = 1 + 2{,}160/\sigma'_{en}$, with σ'_{en}: elastic limit of longitudinal reinforcements (in bar).

Types of columns	Interior columns	Facade columns	Corner columns
Value of θ_1	1	1.4	1.8

Table 8.10. *Values of coefficient θ_1*

Hypothesis	Free structure at one end and recessed at the other	Structure hinged at both ends	Structure embedded at both ends with possible horizontal movement	Structure embedded at one end and hinged at the other	Structure embedded at both ends without the possibility of movement	Braced column of a multi-level building embedded in a foundation block or assembled to floor beams of the same inertia	In other cases
Value l_c	$2\ l_0$	l_0	l_0	$l_0/\sqrt{2}$	$l_0/2$	$0.7\ l_0$	$0.9\ l_0$

Table 8.11. *Buckling length values l_c*

The value of the buckling length lc was defined according to the end connections, as shown in Table 8.11.

It was customary to take for the transverse reinforcements a diameter equal to a third of that of the longitudinal reinforcements and to space them 15 times this diameter. When the longitudinal reinforcements had a minimum and maximum diameter given by $\boldsymbol{\Phi}_{lmin}$ and $\boldsymbol{\Phi}_{lmax}$, the spacing of the transverse reinforcements was given by:

$$\text{Max}\ ((100\Phi_t - 15\Phi_{lmax})\ (2 - \sigma'_b/\sigma'_{b0\ adm};\ 15(2 - \sigma'_b/\sigma'_{b0\ adm})\ \Phi_{lmin}))$$

with:

- Φ_t: diameter of transverse reinforcements;
- Φ_{lmax}: maximum diameter of longitudinal reinforcements;
- Φ_{lmin}: minimum diameter of longitudinal reinforcements;
- σ'_b average compressive stress of the concrete.

Special case of short structures: for structures with a slenderness of less than 50, buckling was not taken into account, and the reinforcement cross-section was given by:

$$N' = \sigma'_b\ (B' + 15A')$$

where B′ is the concrete cross-section and A′ is the cross-section of the longitudinal reinforcements. Cross-section A′ had to be greater than the minimum section A'_m.

8.6.2. *Bending–compression with bending*

8.6.2.1. *Simple bending without compressed steels*

If we consider a rectangular cross-section (without compressed steel) subjected to bending stress under the effect of a moment M, the stress diagram is presented in Figure 8.1.

The cross-sections of the tension reinforcement A were determined from a reduced moment:

$$\mu = 15M/(bh^2\,\overline{\sigma}_a)$$

and:

$$A = M/(\varepsilon\ h\,\overline{\sigma}_a)$$

with:

– $\varepsilon = 1 - \alpha/3$;

– $\alpha = 15/(15 + k)$;

– $k = \sigma_a/\sigma'_b$.

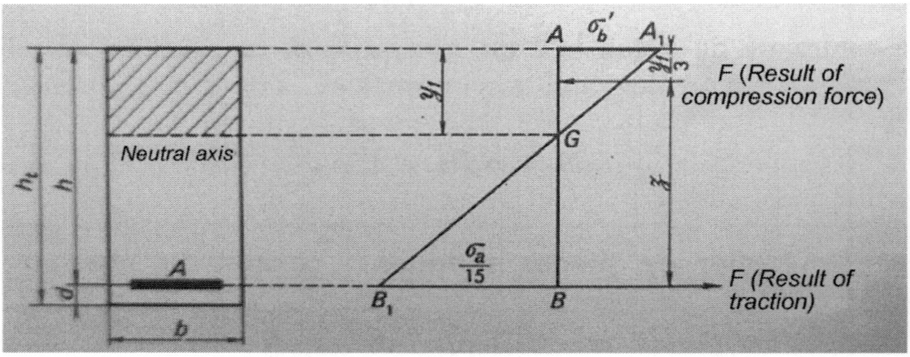

Figure 8.1. *Pure bending stress diagram*

8.6.2.2. *Simple bending with compressed steels*

Prerequisite: compressed reinforcements could only be considered in the calculation of sections if they were tied every 15 Φ according to article 36 of CCBA 68. It was then possible to consider compressed steels in the calculation of the resisting moment according to the diagram in Figure 8.2.

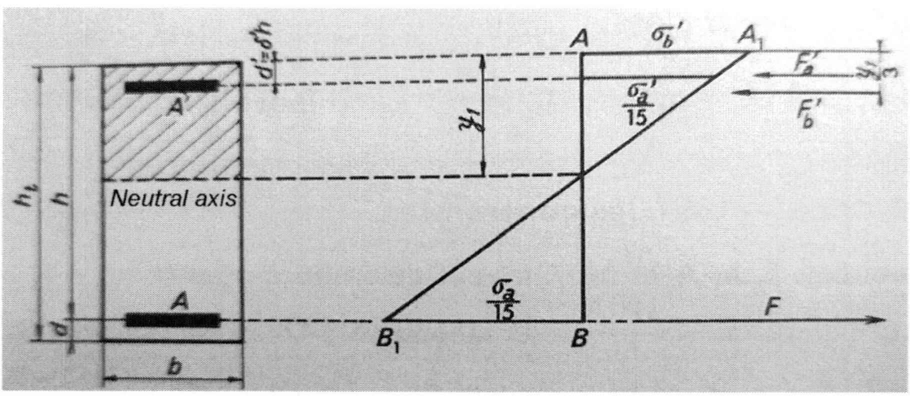

Figure 8.2. *Diagram of the cross-section stresses*

The reinforcement sections A and A′ were then given by:

$$A' = \Delta M/[(h - d')\sigma'_a] \text{ and } A = \acute{\omega} bh/100 + \Delta M/[(h - d')\sigma_a]$$

where $\sigma_a = k\ \overline{\sigma}'_b$ and $\sigma'_a = 15(\alpha - \delta')/\alpha\ \overline{\sigma}'_b$.

8.6.3. *Traction*

Like the structures subjected to simple compression forces, the tensioned structures had to include longitudinal and transverse steels (the latter being reinforced at the overlapping zones of the longitudinal reinforcements). The cross-section A of the longitudinal reinforcements was given by:

$$A = N/\overline{\sigma}_a$$

where N is the normal tensile force and σ_a is the allowable stress in the steel.

The tensile stress of the concrete is given as:

$$\sigma_b = N/(B + 15A)$$

where B is the concrete cross-section and A is the cross-section of longitudinal steels; n = 15.

NOTE.– Special case of circular tanks: if we consider a circular tank of radius R whose walls are subjected to a water pressure p, the tensile force in a vertical section of 1 m of wall was given by:

$$F = pR$$

The cross-section of the horizontal hoops per linear meter was: $A = pR/\overline{\sigma}a$.

The cross-section of the vertical distribution reinforcements (per linear meter of wall) was generally between A/2 and A/3.

8.6.4. *Punching stress*

The punching stress in the thickness of the structure was obtained by the following formula:

$$\tau = 1.5P/(1.2\ p_c\ h_0)$$

with:

– P: punching load;

– p_c: perimeter of the contour parallel to that of the surface along which the load P acts on the slab and is located at the mid-height of the latter;

– h_0 = slab thickness.

8.6.5. *Buckling*

– *Case of axially loaded structures*

The buckling of compressed structures must be considered from the moment when the slenderness of the structure is greater than 50.

The calculation of structures with a slenderness greater than 150 was not covered by the regulations.

For example, $\lambda = l_c/i \geq 50$, where l_c is the buckling length defined in section 8.6.1 and i is the radius of gyration of the structure.

The composite bending structure was then calculated under the effect:

– of an applied normal force N;

– of a bending moment $M = N.f_c$ with $f_c = 8i^2/v\ (1 + \xi)\ 10^{-3}\ (\lambda - 50)^{3/2}$.

In this formula, v denotes the maximum ordinate of the contour of the cross-section of the most compressed side relative to the central axis of inertia of the concrete alone, and ξ denotes the ratio of the permanent load to the total load.

– *Case of structures loaded eccentrically in the buckling plane*

If the slenderness is less than 35, the structure was justified in composite bending without taking into account the transverse strain.

If the slenderness is between 35 and 50, the calculation is carried out with an eccentric normal force of:

$$f_{le} = 0.16\ (\lambda - 35)\ e$$

where e denotes the eccentricity of the load relative to the center of gravity of the concrete cross-section alone.

If the slenderness is between 50 and 150, then a fictitious eccentricity is considered:

$$\lambda'^2 = \lambda^2 + 33{,}000/(1 + 0.6\xi)\ e/h_t$$

where λ denotes the actual slenderness and h_t is the diameter parallel to the buckling plane of the cross-section contour.

NOTE.– Section 33.3 allowed the use of limit states in buckling calculations. In this case, the safety factor between the ultimate load calculated at the limit states and the service load under first-order stress must be greater than 3.

8.7. Constructive provisions

8.7.1. *Concrete pouring*

The possibilities of correct concreting are mentioned in article 44 of the rules. In addition to the provisions described in section 8.4, the following requirements must be respected:

– $r_g \geq C_g/1.4$ if the large granular elements are rounded,

– $r_g \geq C_g/1.2$ if the large granular elements are crushed,

with:

– r_g: ratio of the area to the inner perimeter of a reinforcement mesh,

– Cg: diameter of the largest aggregate.

The need for the installation of concrete pouring and vibration stacks was also mentioned.

8.7.2. *Distance between reinforcements and between reinforcements and formwork*

The stresses provided for CCBA 68 were as follows:

– the cover of each bar must be at least equal to its diameter;

– the maximum diameter of the bars used in the panels and shells must be e/10 (e being the thickness of the panel or shell);

– for two parallel isolated bars, the center distance of the bars must be greater than Max (4 Φ (Φ being the nominal diameter of the bar), 6/5 C_g (C_g being the diameter of the rolled aggregate), and 7/5 C'_g (C'_g being the diameter of the crushed aggregate)). These values were valid when the bars were arranged on a single bed; otherwise, the coefficients 6/5 and 7/5 were increased to 7/5 and 8/5, respectively;

– the plane of the axes of a set of two bars with a diameter greater than 10 mm must be vertical. Sets of two maximum bars should be used, spaced by the nominal diameter of the largest of the bars.

8.7.3. *Concrete joint*

Article 47 specified the following constructive provisions:

– any joint surface that is not fully compressed must be supported by steel reinforcements;

– the allowable tensile stress of the steel reinforcements is equal to one-third of the elastic limit under first-order balanced stresses;

– the steel reinforcements are distributed over the entire joint surface.

8.7.4. *Concrete setting*

The regulations do not provide any information on this subject.

8.8. Comparison between the 1960 regulation (revised in 1964) and that of 1968 (revised in 1970)

Articles	1960 regulation	1968 regulation	Comments
Load combinations	First order: (G) + (T) + 1.2(P) or (G) + (T) + (P) + (V) Second order: (G) + (T) + 1.5(P)	First order: (G) + (T) + 1.2(P) or (G) + (T) + (P) + (V) Second order: (G) + (T) + 1.5(P) + 1,5(V) (G) + (T) + (P) + γ_w(W) (G) + (T) + (P) + (SI) Service: (G) + (T) + (P) + (V)	

Condition of non-fragility	The force required to exceed the elastic limit of the reinforcement must be greater than the breaking force of the uncracked and unreinforced concrete.	Ditto 1960 In 1970, modification of article 19 to avoid unnecessarily increasing the reinforcement of solid structures	
Compressive strength of concrete	Measured from 16 × 32 test tubes at 28 days σ_{28} or σ_n	Measured from 16 × 32 test tubes at 28 days σ'_{28} or σ'_n	
Allowable compressive stress of concrete	$\sigma_b = \alpha\beta\gamma\delta\varepsilon\ \sigma_{28}$	$\sigma'_b = \alpha\beta\gamma\delta\varepsilon\ \sigma_{28}$	
Allowable tangent stress	When the compressed zone has the same width as the core b_0 (rectangular cross-section) $T_b = T/(b_0Z) \leq 2.3\,\bar{\sigma}'_b$ When the average width b_m of the compressed zone is greater than two times the core width b_0 $T_b = T/(b_0Z) \leq 2.9\,\bar{\sigma}'_b$	If the concomitant compressive stress of the concrete σ'_b is less than the permissible stress of concrete in compression $T_b = T/(b_0Z) \leq 2.3\,\bar{\sigma}'_b$ Otherwise $T_b = T/(b_0Z) \leq (4.5 - \sigma'_b/\sigma'_{b0})\,\bar{\sigma}_b$	
Concrete modulus of elasticity	Instant module: $Ei = 21{,}000\ \sqrt{\sigma_j}$ (bars) Deferred module: $Ev = 27{,}000\ \sqrt{\sigma_j}$ (bars)	Instant module: $Ei = 21{,}000\ \sqrt{\sigma'_j}$ (bars) Deferred module: $Ev = 27{,}000\ \sqrt{\sigma'_j}$ (bars)	In 1968, for $j \geq 28$ days, the value of σ'_j was $1.20\sigma'_{28}$ for class 325 cement and $1.10\sigma'_{28}$ for other types of cements
Allowable stress of steel in tension/compression	$\bar{\sigma}'_a = 2/3\ \sigma'_{en}$ (without exceptions)	$\bar{\sigma}_a = 2/3\ \sigma_{en}$ (without exceptions)	
Principles of structural calculations	Calculation in allowable stress Load values taken at their maximum	Ditto 1960 with the possibility of working at limit states	

Table 8.12. *Design rules*

Articles	1960 regulation	1968 regulation
Steel cover	– 40 mm for seafront structures – 20 mm for other structures – 10 mm for interior structures in non-condensing premises	Ditto 1960
Distance between steels	– Distance to allow the implementation and tightening of the concrete between the reinforcements – The minimum free horizontal distance between two bars of the same layer was worth 6/5 Cg (where Cg designated the maximum dimension of the largest aggregate) – The vertical distance between a pack of two bars and the third was worth: Max (3/4Φ; 1/2 Cg)	– The cover of each bar must be at least equal to its diameter – The maximum diameter of the bars used in the panels and shells must be e/10 (e being the thickness of the panel or shell) – For two isolated parallel bars, the distance between the bars must be greater than the Max (4 Φ (Φ being the nominal diameter of the bar); 6/5 Cg (Cg being the diameter of the rolled aggregate); 7/5 C′g (C′g being the diameter of the crushed aggregate). These values were valid when the bars were arranged on a single bed, otherwise the coefficients 6/5 and 7/5 were increased to 7/5 and 8/5, respectively – The plane of the axes of a bundle of two bars with a diameter greater than 10 mm must be vertical. It is advisable to use bundles of two bars maximum spaced by the nominal diameter of the largest of the bars
Arrangements of reinforcements for slabs and joists	Rules for attachment seams in areas of overlap, anchoring, etc.	Ditto 1960 plus minimum values of the "mesh"

Table 8.13. *Constructive provisions. Comparison of the BA 60 and CCBA 68 regulations*

8.9. Example of calculating an underground tank

We consider a semiburied cylindrical tank 10 m in diameter and 8 m in height, according to the diagram in Figure 8.3.

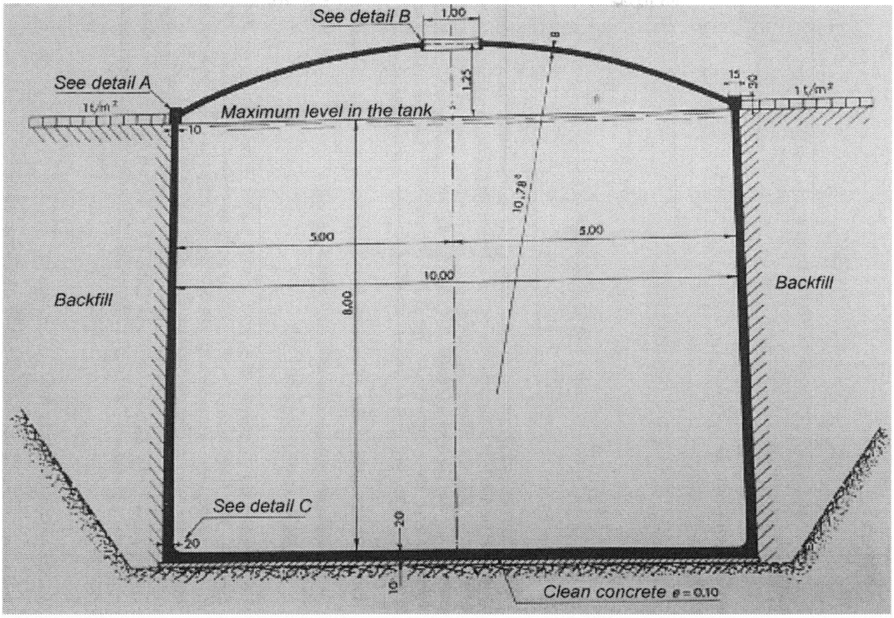

Figure 8.3. *Tank cross-section plan*

The sizing assumptions are as follows:

1) Squaring of structures:

- reinforced concrete cover, 8 cm thick with self-protected waterproofing cover;

- vertical walls of variable thickness from 0.20 m at the base to 0.10 m at the top;

- reinforced concrete foundation slab 0.20 m thick.

2) Applied loads:

- the density of the liquid contained is equal to 1 t/m^3. The water level in the tank is 8 m;

- an operating load of 1 t/m^2 is considered on the back fill at the periphery of the basin.

3) Materials:

- the concrete making up the structure is dosed at 400 kg/m^3 and is not very controlled;

- the frames are made of mild steel FeE22;
- the stress on the ground is limited to 2 bar.

The checks were then carried out as follows:

1) *Verification of ground stress (Table 8.14)*

Structural element	Load calculation	Result
Covering dome	Concrete weight: 0.08 × 2,500 × 2 × π × 10.826 × 1.25 = 16,997 kg Cover weight: 250 × 2 × π × 10.826 × 1.25 = 21,246 kg	38,243 kg
Upper tank belt	0.30 × 0.15 × π × 10.10 × 2,500 = 3,568 kg	3,568 kg
Tank walls	(0.10 + 0.20)/2 × 8 × π × 10.10 × 2,500 = 95,142 kg	95,142 kg
Tank base	0.20 × π × 10.40^2/4 × 2,500 = 42,453 kg	42,453 kg
Liquid contained	8 × π × 10^2/4 × 1,000 = 628,000 kg	628,000 kg

Table 8.14. *Breakdown of loads*

The total weighted load at ground level is then G + 1.2Q, that is, 179,406 + 1.2 × 628,000 = 933,000 kg. Let there be a ground stress of $\sigma' = 933{,}000/(\pi \times 10.40^2/4)$ = 1.10 bar ≤ 2 bar.

2) *Checking the cover dome*

The weight per linear meter of dome is given by p = 38,243/(2 × π × 5.05) = 1,206 kg/lm. This is a normal force in the dome of 1,206 × 10.826/5.05 = 2,585 kg.

The thrust at the base of the dome at the level of the upper tank belt is therefore: H = 1,206 × (10.826 – 1.25)/5.05 = 2,287 kg/lm.

The compressive stress in the concrete of the dome is then given by: σ'_b = 2,585/(8 × 100) = 3.23 bar. This value is compared with the allowable stress for concrete with attenuated control and a dosage of 400 kg/m^3, that is, 7.5 MPa (75 bar).

A minimum percentage of reinforcements was then implemented corresponding to:

– 8ϕ8/lm along the meridians;

– 5ϕ6/lm along the parallel direction.

3) *Checking the upper tank belt*

The tensile forces in the belt are given by: T = 2,287 × 5 = 11,435 kg. Let us take a cross-section of steel A_s = 11,435/1,000 = 11.44 cm² (6ϕ16 = 12.06 cm²).

4) *Checking the tank*

8.9.1. *Full tank*

NOTE.– The actions to be considered in this case for dimensioning the tank walls are:

– the thrust of water;

– ground thrust;

– the thrust of variable loads on the central reservation.

While it was common practice not to consider the loads on the earthwork to oppose the thrust of the internal water of the structure, it was common in the literature to consider the thrust of the earth (partial or total) to reduce the pressure of the water on the wall. *This arrangement led to underestimating the tensile forces in the wall and the quantity of steel required.*

Section	Water thrust p (kg/m²) considered in the center of the section	Traction in the concrete wall (weighted by 1.20) T = pR (kg)	Required steel cross-section (cm²)
From 0 to 1 m	500	3,000	3
From 1 to 2 m	1,500	9,000	9
From 2 to 3 m	2,500	15,000	15
From 3 to 4 m	3,500	21,000	21
From 4 to 5 m	4,500	27,000	27
From 5 to 6 m	5,500	33,000	33
From 6 to 7 m	6,500	39,000	39
From 7 to 8 m	7,500	45,000	45

Table 8.15. *Traction force in the walls under water pressure*

The water pressure is calculated per 1 m section of wall, and the forces are summarized in Table 8.15.

NOTE.– If the ground pressure of 50% was taken into account to counterbalance the water pressure, the reinforcement cross-section in the 6–7 m and 7–8 m sections then became 28 cm² and 32 cm², respectively, an underestimation of almost 30%.

8.9.2. *Empty tank*

The ground pressure on the walls of the reservoir is given for a wet earth density of 2,000 kg/m³ and a thrust coefficient of 0.27, that is, $P_1 = 0.27 \times 2{,}000 \times 8 = 4{,}320$ kg/m².

The variable loads on the ground (Caquot loads of 1 t/m²) generate pressure on the wall $P_2 = 0.27 \times 1{,}000 = 270$ kg/m². According to CCBA 68, this variable load is weighted by 1.20, that is, a calculation value of $1.20 \times 270 = 320$ kg/m².

These efforts generate compressive stresses in the wall. These values are determined as previously described by cutting the wall into 1 m high sections.

Section	Total ground thrust (kg/m²)	Compression within the wall (kg)	hm (cm)	A′ (cm²)	B′ = 100 hm + 15A′	σ′ = F′/B′ (kg/cm²)
From 0 to 1 m	590	2,950	10.6	3	1,105	2.7
From 1 to 2 m	1,130	5,650	11.9	9	1,325	4.3
From 2 to 3 m	1,670	8,350	13.1	15	1,535	5.4
From 3 to 4 m	2,210	11,050	14.4	21	1,755	6.3
From 4 to 5 m	2,750	13,750	15.6	27	1,965	7
From 5 to 6 m	3,290	16,450	16.9	33	2,055	8
From 6 to 7 m	3,830	19,150	18.1	39	2,395	8
From 7 to 8 m	4,370	21,850	19.4	45	2,615	8.4

Table 8.16. *Compressive force in walls under ground thrust and live loads*

The concrete is poorly controlled and dosed at 400 kg/m³. The allowable compressive stress is therefore 75 bar.

8.10. Earthquake Regulations of 1969

8.10.1. *Overview*

Published in 1969, the PS 1969 regulations were intended to ensure a certain protection of people and property against the effects of earthquakes.

These regulations, which took on the status of DTU, came after:

– AS 55 recommendations published following the Orléansville (Algeria) earthquake in 1955;

– PS 62/64 regulations drawn up following the Agadir (Morocco) earthquake in 1964.

To this end, the Instruction therefore included, alongside general considerations, recommendations whose application was strongly advised.

They were based on the following principles:

– the concept of nominal protection for seismic intensity i_N;

– the choice of the nominal earthquake intensity i_N.

The choice of calculation methods and the value of the coefficients then made it possible for nominal protection corresponding to a given intensity to ensure that:

– there is a low probability of the occurrence of structural damage requiring repair;

– there is a very low probability that the uncollapsed building will not be repairable;

– there is an even lower probability that the building will collapse.

The seismic zoning of France is presented in Figure 8.4.

The territories were divided into four categories:

– Zone 0: no seismicity;

– Zone 1: low seismicity;

– Zone 2: average seismicity;

– Zone 3: high seismicity.

The structures were classified according to their importance, as shown in Table 8.17.

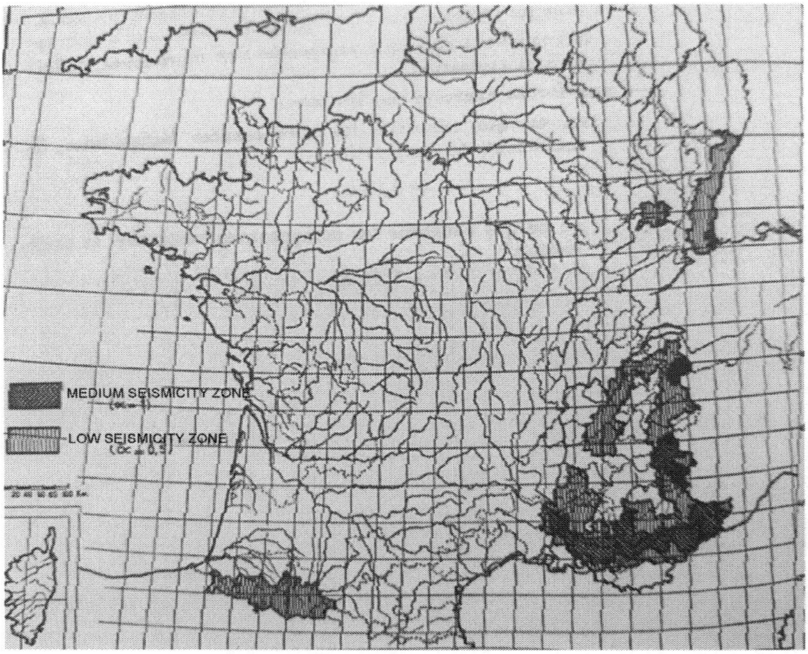

Figure 8.4. *Seismic zoning of France in 1969*

Group	Definition	Examples	i_N value Zone 1	i_N value Zone 2	i_N value Zone 3
I	Buildings posing a so-called normal risk to the population	Homes, offices, factories, workshops	7 ($\alpha = 0.5$)	8 ($\alpha = 1$)	8.6 ($\alpha = 1.5$)
II	Buildings presenting a high risk due to their attendance, their importance	Schools, stadiums, performance halls, train station halls, thermal power stations	7.6 ($\alpha = 0.75$)	8.3 ($\alpha = 1.2$)	8.8 ($\alpha = 1.7$)
III	Structures whose safety is essential for civil protection needs	Hospitals, barracks	8 ($\alpha = 1$)	8,6 ($\alpha = 1.5$)	9 ($\alpha = 2$)
IV	Structures whose disorganization presents a particularly serious risk	Nuclear power plants, etc.	To be defined by administration	To be defined by administration	To be defined by administration

Table 8.17. *Table of nominal intensity values according to group importance*

8.10.2. *Principles of seismic design*

The 1969 rules applied to all constructions, with the exception of civil engineering structures, for which protection against earthquakes was required either by public regulations or specifications.

The recommendations attached to the Instruction stressed the following points:

– the need to design simple structures with regular, symmetrical shapes. The distribution of masses in height was avoided as well as "open" floors (transparent);

– the need to provide joints to split the construction into simple and regular shapes. These joints had to be flat and "clean". The width of the joints had to be greater than 2 cm in all the cases;

– at the foundation level, the base soil had to be chosen from compact and non-compartmentalized land, in preference to fractured soils, alluvial soils, or those with a high void index or saturated with water. Deep foundations were recommended;

– the need to connect the foundations to each other and to the superstructure. The aim was to reduce relative movements between foundations;

– the need to connect internal and external pipes (gas, water, etc.) to avoid breakages;

– at the superstructure level, it was advisable to avoid significant overhangs, particularly because of their sensitivity to the vertical earthquake component.

Bracing is the most critical part of earthquake-resistant design.

The recommendations drew the designer's attention to the importance of longitudinal bracing as well as transverse bracing.

Symmetrical bracing was recommended (to avoid torsional oscillations).

Building restrictions:

– ribbed reinforced concrete floors without a compression slab;

– ceilings made of fragile materials (clay, plasterboard, etc.);

– vaults without tension rods;

– stairs built into a wall or of the Saracen vault type.

8.10.3. *Special regulations for reinforced concrete structures*

In addition to the constructive provisions of the CCBA 68 regulations, PS 69 requested the following:

– At the assembly nodes, a concrete hooping over a distance equal to $b_0 + h$ (b_0 width of the beam and h its height). The transverse reinforcements had to connect all the longitudinal rows (buckling). In the columns, on either side of the node over a length equal to 1/6 of the free height of the column, the maximum spacing of the frames had to be reduced to 4/5 of the normally calculated space.

– At the columns, hooks were prohibited at the ends and at the overlaps of the longitudinal reinforcements. The minimum overlap length was 50ϕ (ϕ being the diameter of the bar) for mild- and high-adhesion steels. For smooth medium-hard round bars, this value was increased to 80ϕ.

– For structures subjected to tensile stresses, the overlap lengths of the longitudinal reinforcements had to be free of hooks, and their values equal 130% of the required value. Frames had to fix the overlap areas (spacing less than twice the smallest structure size), and their cross-section had to be greater than 130% of the normal cross-section.

– For the transverse reinforcements of the prismatic structure subjected to a punching force, the spacing of the frames must be less than the useful height of the structure.

– At floor level, the ribbed reinforced concrete floors had to have a compression slab of at least 4 cm.

8.10.4. *Calculation method of reinforced concrete structures*

The calculation of the structures was initially carried out under static stresses, and then the structures were verified under seismic stresses. The justification was based on:

– overall strength and stability;

– taking into account deformations in the verification;

– checking the stability of certain elements.

The loads taken into account in these checks are summarized in Table 8.18.

To the stresses resulting from the loads in Table 8.18, the following was appropriate:

– seismic stresses;

– 4/5 of the operating load of residential buildings does not generate seismic effects but takes into account gravity loads;

– the effects of temperature and shrinkage;

– stresses resulting from the relative horizontal displacement of the foundations of long buildings.

Load type	Residential buildings, offices, hospitals, etc.	Buildings open to the public (rooms, stores, libraries, etc.), garages, workshops, industrial buildings (except storage areas), etc.	Warehouses, storage rooms, tanks, silos, etc.
Permanent loads	X	X	X
Snow load beyond 35 daN/m²	X	X	X
Operating loads	1/5 of operating loads (without degression)	Total loads due to machinery and equipment (except suspended ones) + 1/2 other operating loads	Total operating loads

Table 8.18. *Combination of loads for calculating seismic stresses*

The resistance of the structural elements was justified according to the following requirements:

– foundations: the stress should not exceed 75% of the foundation's resistance capacity for collapse;

– structures: the allowable stresses are summarized in Table 8.19.

NOTE.– When calculating the strength of reinforced concrete elements, the only vertical elements that could be considered to balance bending stresses were those with a cross-section greater than 225 cm² and a useful height greater than 13 cm.

Structural analysis	Concrete: permitted stress	Reinforcements: permitted stress
Limit states	Limit state of resistance not reached	Limit state of resistance not reached
Allowable stresses	Compression: 85% of the resistance at 28 days measured on cylindrical specimens	Elastic limit of steel

Table 8.19. *Allowable stresses*

8.10.5. *Structural calculation methods*

The regulations defined two methods of seismic calculation:

– Equivalent static calculations: the principle was to substitute dynamic stresses for static stresses from static force systems whose effects were equivalent to those of seismic forces. This method was largely developed in 1969 regulations.

– Direct calculations: the constructors could justify their building using any scientifically valid method of their choice. In particular, justifications based on direct calculations from accelerograms of tremors recorded a short distance from the epicenter.

8.10.5.1. *Equivalent force methods*

In the case of this method, the equivalent force systems were as follows:

– Horizontal force systems (S_H) composed of elementary forces, each acting on a construction element and applied to its center of gravity. The intensity of each elementary force was: σ_h W, where σ_h is the "seismic coefficient of the element for the horizontal direction" and W is the weight of the loads specific to this element.

– Vertical force systems (S_V) composed of elementary forces each acting on a construction element and applied to its center of gravity. The intensity of each elementary force was: $\pm\sigma_v$ W, where σ_v is the "seismic coefficient of the element for the vertical direction" and W is the weight of the loads specific to this element.

– Overall torsion (S_T): this stress had to be considered when the thrust η of the flat cross-section was greater than 2.5 or when more than two-thirds of the horizontal forces were balanced by the same bracing plane. In this case, an eccentricity was defined:

$$e = 1/20\ (\lambda + \mu)\ (L_y + d)$$

In this formula, L_y represents the plane of the level perpendicular to the direction of the horizontal forces considered, d is the lever arm of the theoretical result of the horizontal forces of the level relative to the center of torsion of the structure, and λ and μ are two coefficients where:

– $\lambda = 0$ if $\eta \leq 2.5$ and $\lambda = \eta - 2.5$ otherwise;

– $\mu = 0$ if the most stressed bracing plane balanced less than two-thirds of the horizontal forces;

– $\mu = 6\ (\theta - 2/3)$ if, in the most stressed bracing plane, a fraction θ of the horizontal forces greater than two-thirds was balanced.

The determination of seismic coefficients depended on the nature of the structure. A distinction was made between common constructions and those with several degrees of freedom in the same horizontal plane or with multiple oscillators with low damping and relative flexibility ($T \geq 0.5$ s).

– In the first case, the seismic coefficient was given by the following formula: $\sigma_x = \alpha\beta\gamma\delta$, where α is the intensity coefficient, β is the response coefficient, γ is the distribution coefficient and δ is the foundation coefficient. The values of these different coefficients are as follows:

- α depended on the nominal intensity i_N. $\alpha = 1$ for $i_N = 8$; for other values, see Table 8.17;

- β characterized the importance of the response of the structure, depending on the period T of the fundamental mode, the degree of damping of the structure and the nature of the soil foundation;

- for normal damping: $\beta = 0.065/(T)^{0.33}$ with a minimum of 0.05 and a maximum of 0.085 for foundations on soft ground and 0.10 for other cases. Normal damping concerned traditional buildings used for housing, offices, etc.;

- for average damping: $\beta = 0.085/(T)^{0.33}$ with a minimum of 0.065 and a maximum of 0.110 for foundations on soft soils and 0.130 for other cases. The average damping concerned large halls, buildings with light partitions, etc.;

- for low damping: $\beta = 0.105/(T^3)^{0.25,}$ with a minimum of 0.075 and a maximum of 0.175 for foundations on soft soils and 0.200 for other cases. The formula was only valid for $T \leq 0.5$ s. Low damping concerned pylons, masts, tanks on towers, etc.;

- γ depended only on the structure and characterized the behavior of the mass and was given by the following formula:

$$\gamma(h) = A(h) \frac{\Sigma M(Z)\, A(Z) + \int m(z)\, A(z)\, dz}{\Sigma M(Z)\, A^2(Z) + \int m(z)\, A^2(z)\, dz}$$

where M(Z) is the mass concentrated at the Z coordinate, m(z) is the mass distributed per unit length at the z coordinate and A(Z) is the elongation in the main mode of oscillation;

- δ was defined according to the nature of the soils (see Table 8.20).

Type of foundation	A Rocky terrain	B Medium consistency soils	C Soft soils, fairly high water content	D Waterlogged silts
Surface footings	1	1.15	1.25	
Footings on wells	0.90	1.00	1.15	
Foundation slab		1.00	1.10	1.20
Point-supported pile		1.10	1.15	1.30
Floating piles		1.10	1.30	

Table 8.20. *Values of the coefficient δ depending on the nature of the ground and the foundation*

NOTE.– The seismic coefficient in the vertical direction was σv = +/–σH/√α.

The values of the period T (s) were the subject of flat-rate formulas for simple-shaped buildings, depending on the height H (m) of the building and the plane Lx (m) in the direction considered.

For bracing by masonry or unreinforced concrete walls: T = 0.06H √H/√ (Lx (2Lx + H)).

For bracing by reinforced concrete walls: T = 0.08H √H/√ (Lx (Lx + H)).

For reinforced concrete frame bracing: T = 0.09H/√Lx.

For metal frame bracing: T = 0.10H/√Lx.

– In the second case, the seismic coefficient was given by the formula: $\sigma_i = \alpha\beta_i\gamma_i\delta$, where *i* represents the *i*th mode.

The values of β_i were modified (see Table 8.21).

Value of T_i (period of the mode considered)	$0 \leq T_i \leq 0.2$ s	$0.2 \leq T_i \leq 0.26$ s	$0.26 \leq T_i$
β_i	1.25 T_i	0.25	$0.09/(T_i^3)^{0.25}$

Table 8.21. *Values of coefficient β_i*

NOTE.– The minimum value of β_i was set at 0.06.

For a structure built on a significant layer of soft soil, the values of $\boldsymbol{\beta_i}$ were modified (see Table 8.22).

The stresses Σ_i were then determined by applying the seismic coefficients σ_1 and σ_v, as well as σ_2 and σ_3, that is,

$$\Sigma = (\Sigma_1 + \lambda_2\Sigma_2^2 + \lambda_3\Sigma_3^2)^{0>5}$$

with:

– $\lambda_2 = 4T_1 - 2$ (with a maximum of 1 if $T_1 \geq 0.75$ s and a minimum of 0 if $T_1 \leq 0.5$ s);

– $\lambda_3 = 4T_1 - 3$ (with a maximum of 1 if $T_1 \geq 1$ s and a minimum of 0 if $T_1 \leq 0.75$ s).

NOTE.– The regulations also made recommendations for:

– structures with several degrees of freedom in the same horizontal plane;

– structures with different degrees of damping at various levels;

– long structures and highly hyperstatic structures.

Value of T_i (period of the mode considered)	$0 \leq T_i \leq 0.16$ s	$0.16 \leq T_i \leq 0.35$ s	$0.35 \leq T_i$
β_i	$1.25\ T_i$	0.20	$0.09/(T_i^3)^{0.25}$

Table 8.22. *Values of coefficient β_i for soft soils*

8.10.5.2. *Direct calculation methods*

The calculation was carried out on the basis of accelerograms recorded at a short distance from the epicenter. It was also necessary to consider vertical components with an intensity at least equal to that of the horizontal tremors. In the case of deformable structures, it was necessary to consider second-order effects.

When calculating the safety verification relating to deformations, considering the potential increase in the latter due to the deformability of the structure was recommended.

8.10.5.3. *Safety against deformation*

Width of expansion joints: the minimum width of the joints was set at *2 cm*. It had to be calculated in such a way as to avoid the collision of neighboring buildings, taking into account the loss of rigidity of the structural elements. With no special justification, the value of the seismic displacement takes the value of the elastic displacement multiplied by 1.15 for panel-braced structures and triangulated beams and by 1.30 for frame-braced structures. These coefficients increased to 1.45 and 1.60 when the oscillation periods were determined from the fixed formulas.

Maximum strain: to avoid damage to fragile non-structural materials (glass facades, plaster, etc.), the relative movement between two consecutive levels was limited to $(\sqrt{\alpha}/1\,000) \times h$ (with h: floor height; α: intensity coefficient).

8.10.5.4. *Special cases*

Consoles and corbels: the cantilevered parts of the structures were dimensioned, taking into account a seismic coefficient applied to all the loads: $\mathbf{0.15\alpha/T}$. The value of this coefficient was between 0.4α and 1. T is the natural period of the oscillator formed by the cantilever element in the vertical direction and α is the intensity coefficient. For verification, only vertical seismic forces were considered.

Arch tie rods, hangers, beams receiving a column: these parts were calculated in the vertical axis under all loads with a seismic coefficient $b_{ui} = +/-2\sigma_H/\delta$, where σ_H represents each load and live load, the largest seismic coefficient taken into account in the horizontal directions.

Walls and partitions: for seismic intensities greater than 7, at their breaking limit, the walls and partitions must be capable of resisting a stress perpendicular to their plane and corresponding to a seismic coefficient of:

– $0.10 + 0.05\alpha$ for walls and partitions;

– 0.20 + 0.10α for prefabricated facade panels, curtain walls and prefabricated partitions;

– 0.40 + 0.15α for various veneers and coverings.

Retaining walls: for the calculation, a seismic coefficient identical to that of the isolated walls of 0.10α was applied to the wall itself.

For the stresses exerted by the ground, the horizontal components were increased by 1 + 0.10α, and the vertical components were increased by 1 ± 0.10α.

8.11. Special cases of silos

8.11.1. *Overview*

As mentioned previously, the subsequent regulations that followed until the publication of CCBA 68 were as follows:

– the Koenen–Janssen formula in 1895;

– the formulas of Messrs. Caquot and Kérisel in 1956;

– German standards DIN 1055-1 and 1055-6 in 1964.

In 1975, in France, "professional guidelines for the design and calculation of concrete silos" appeared. This document produced by SNBATI was published in *Annales de l'ITBTP* in December 1975 under the No. TMC/189. These guidelines set the characteristics of a silo, the method of calculating the pressures exerted on the walls and the values of the parameters for calculating the interaction between the ensiled material and the wall. They were limited to powdery products and excluded all those that had or could acquire a certain cohesion.

They were modified in 1986 in a second edition in the following areas:

– modification of the methods for determining the ovalization moments and the thermal gradient;

– use of BAEL 83 rules to replace CCBA 68 rules;

– appearance of prestressed concrete silos.

The silage material was always characterized by:

– its density: γ;

– its internal friction angle: φ;

– its friction angle on the concrete wall: δ.

Material	Density γ in kN/m^3	Internal friction angle φ in degrees
Cement	14.7	28
Clinker	14.7–15.7	33
Cement raw material	13	26
Gypsum	12.25	25
Coal powder	8.35	25
Wheat	8.35	26
Corn	7.85	24
Barley	8.1	24
Rapeseed	7.0	24
Soybean seeds	7.5	27
Sunflower	5.0	22
Sorghum	7.35	23
Hulled rice	8.6	30
Wheat flour	8.8	20
Sugar	9.0	30

Table 8.23. *Volumetric weights and internal friction angles of silage materials*

The values are summarized in Table 8.23.

The walls were classified into three categories according to their roughness:

– wall A: wall with horizontal grooves;

– wall B: uncoated concrete;

– wall C: concrete coated with paint.

The conventional value of the material/wall friction angle was defined by the relation:

$$\rho = tg\delta / tg\varphi$$

The values of ρ were taken from Table 8.24.

Type of material	Examples	Wall A	Wall B	Wall C
Powders	Cement, plaster, flour, etc.	0.87	0.80	No value
Small grains	Wheat, corn, barley, sorghum, sugar, etc.	0.87	0.75	No value except for sugar (0.50)
Large grains	Clinker	0.87	0.70	No value
Special cases	Rapeseed, soybeans, sunflower, rice, etc.	0.75	0.65	No value

Table 8.24. *Values of ρ*

The recommendations used the following terminology:

– Hydraulic radius: $r_H = S/L$, where S is the area of the cross-section between the outer and inner walls. The most common values of the hydraulic radius are rH = r/2 for a circular section of radius r or a regular polygon with an inradius r, $r_H = ab/(a + b)$ for a rectangular section with short side 2a and long side 2b, $r_H = (r - r')/2$ for a circular crown section of large radius r and small radius r′.

– Geometric data: see Figure 8.5.

– Slenderness: the slenderness of a storage structure is the H/r_H ratio. The limit values in the recommendations were:

- 3.5 for the slenderness ($H/r_H \geq 3.5$),

- 7.50 m for the hydraulic radius of silos with normal, geometrically abnormal, mechanically abnormal emptying and for structurally abnormal emptying with a central chimney if the silage material was a cereal,

- 15 m for the hydraulic radius of silos with structurally abnormal emptying (with numerous openings and a low flow rate) or those whose ensiled material is sugar or with a slenderness less than 6;

– normal emptying: without any structure inside and entirely contained inside a centered contour. The silo emptied from the bottom by gravity from the silo cross-section by homothety with a ratio of 0.4;

– geometrically abnormal emptying: this was the case where the first two conditions of normal emptying were met but not the third;

– mechanically abnormal emptying: this is the case where the silo emptied through the bottom or through a sidewall with the insufflation of pressurized air;

– structurally abnormal emptying: this was any type of emptying that involved the internal structure of the silo or programed emptying and filling processes (emptying by chimney, emptying by low flow openings located on a foundation slab, etc.).

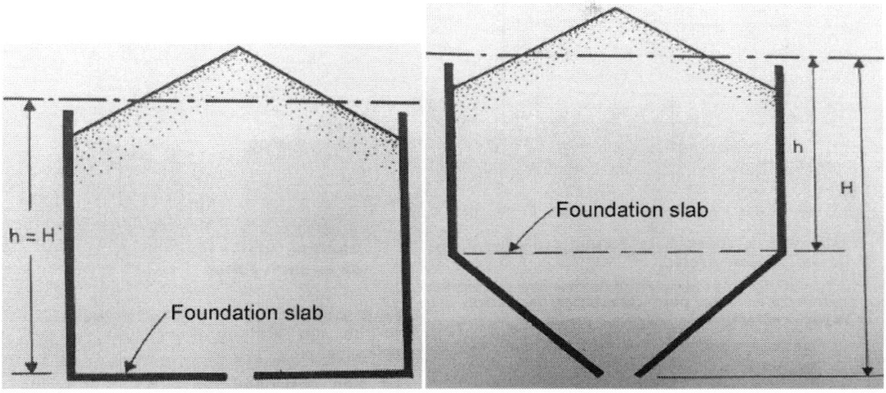

Figure 8.5. *Geometric data*

8.11.2. *Determining actions on the vertical walls and the bottom of a silo*

The actions were defined in nominal values on the basis of two equilibrium limit states. These nominal values were obtained by multiplying the basic actions by a behavior coefficient k. Each of the limit states was characterized by a value of the parameter λ. The first equilibrium state was that which occurred after filling when the latter was carried out in a single phase and when the powder mass was at rest. This was generally the state that determined the actions on the bottom of the silo. The second equilibrium state concerned the actions that occurred at the beginning of emptying or at any time during storage. This state generally determined the dimensions of the sidewalls.

Average vertical pressure on a horizontal plane was defined by v_0 (pressure exerted by the ensiled material on the horizontal plane at depth z relative to the average filling plane) such that:

$$v(z) = n(z)/\lambda + \gamma\, h''$$

In this formula, λ is a parameter that takes the values:

– $\lambda_1 = (1 - m \sin\varphi) \cos^2\delta/(1 + m \sin\varphi)$ for the first equilibrium state (with $m = (1 - \rho^2)^{1/2}$);

– $\lambda_2 = \cos^2\delta$ for the second equilibrium state.

The value of h″ was given by:

– $h'' = 1/2\ r_H\ tg\delta$ for silos with regular circular or polygonal cross-section;

– $h'' = a/8\ (3 - a/b)\ tg\delta$ for silos with a rectangular cross-section and short side 2a and long side 2b.

n(z) designated the value of the horizontal component of the thrust on the vertical wall at the same depth z.

The horizontal pressure n(z) on the vertical wall was defined by:

$$n(z) = (\gamma\ r_H/tg\delta)\ y$$

So if:

$$n(z) = \gamma\lambda z_0\ (1 - \exp\ (-(z - h'')/z_0))$$

z_0 is the reference height defined by $z_0 = r_H/(\lambda\ tg\delta)$.

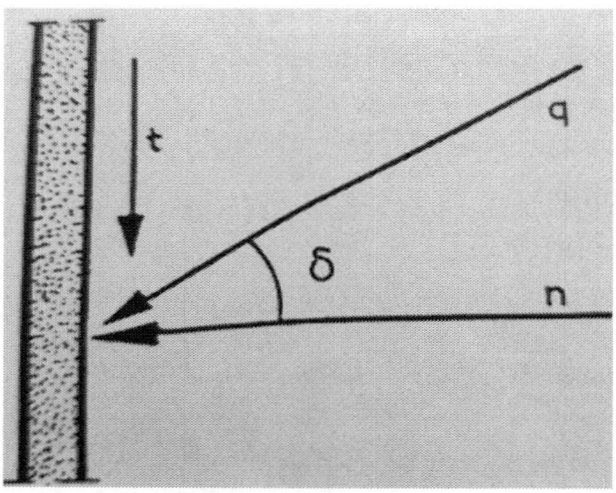

Figure 8.6. *Distribution of stresses to the right of the wall*

The vertical pressure t on the vertical wall was given by:

$$t(z) = n(z)\ tg\delta = \gamma\ r_H\ (1 - \exp\ (-(z - h'')/z_0))$$

8.11.3. *Determining nominal actions on the vertical walls and the bottom of a silo with normal emptying*

Nominal actions on the silo structures were established from the previous definitions after taking into account behavior coefficients k_n and k_v, whereby:

– $n = k_n (\gamma\ r_H/tg\delta)\ y$ (nominal value of the horizontal action of the silage material on the vertical wall);

– $t = n.tg\delta$ (vertical action of the silage material on the vertical wall);

– $\overline{v} = k_v\ \gamma\ (z_0.\ \overline{y} + h'')$ (nominal value of the average vertical action exerted by the ensiled material on a horizontal plane at depth z relative to the average filling plane);

– $T = \gamma.r_H.z_0.\ (x - y)$ (resultant of friction forces on a unit width strip over the entire height of the silo above the base plane).

The values of the coefficients k_n and k_v were as follows:

– $k_n = 1.15$ (this coefficient integrates the excess pressures that may appear on the wall of the silo);

– $k_v = 1.35$.

8.11.4. *Determining nominal actions on the vertical walls and the bottom of a silo with abnormal emptying*

8.11.4.1. *Determining nominal actions on the vertical walls and the bottom of a silo with geometrically abnormal emptying*

The second equilibrium state was replaced by the next one (state 3) according to the notations in Figure 8.7.

In silos with a rectangular cross-section, the normal actions on the walls were increased according to the following formulas:

$$n_3 = n_2 + \overline{n}_1/15.\ (1 - 4u/5Ht) \text{ if } 0.6 \leq \xi \leq 1$$

$$n_3 = n_2 + \overline{n}_1/15.\ (1 - 4u/5Ht)\ (5\xi - 2) \text{ if } 0.4 \leq \xi \leq 0.6$$

with:

– $n_2 = k_n (\gamma\ r_H/tg\delta)\ y$;

– $\overline{n}_1$: value of n_1 determined with $z = h$ and $y = \overline{y}$;

– ξ: relative coordinate according to the diagrams in Figure 8.8;

– u and Ht are determined in Figure 8.7.

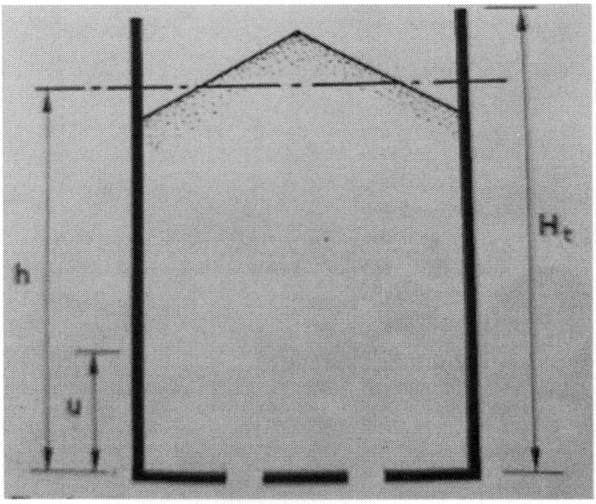

Figure 8.7. *Geometric dimensions*

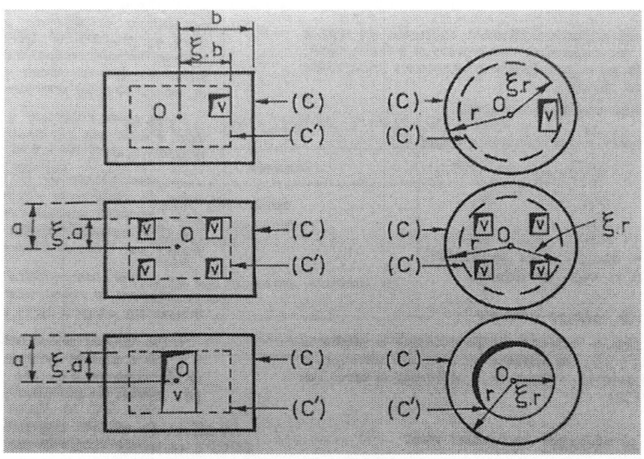

Figure 8.8. *Relative coordinates*

NOTE.– Compared with a silo with geometrically normal emptying, the forces increase by a ratio of n_3/n_2.

In silos with a circular cross-section (or a polygon with many sides), it was also appropriate to consider when determining the dimensions of an ovalization moment according to the following different provisions:

– sleeve attached at the foot to the floor or a rigid belt preventing ovalization at this level but free at the head;

– sleeve attached at the foot and top of a floor or belt preventing ovalization at these levels;

– free sleeve at the foot (connection at the head not included).

The values of the ovalization moment depended on the previous configuration and the area in which the stress occurred. Thus, for the wall, the ovalization moment was as follows:

$$M_{ov} = \pm M_{ref}.G(\psi.u/H_t).1/(1 + (4.n_{02}.r^3)/(E_i.h_0{}^3))$$

For the head belt, it became:

$$M_{ovt} = \pm M_{ref}.H_t.K_t(\psi)$$

For the bottom belt, it became:

$$M_{ovp} = \pm M_{ref}.H_t.K_p(\psi)$$

In these formulas:

– M_{ref} is the reference moment defined by:

$$M_{ref} = \mu\,\overline{n}_1\ r^2 \text{ if } 0.6 \le \xi \le 1$$

$$M_{ref} = \mu\,\overline{n}_1\ (5\xi - 2)\, r^2 \text{ if } 0.4 < \xi < 0.6$$

– r is the radius of the silo;

– the coefficient μ depends on the silo type (see Table 8.25);

– u is the height of the point considered;

– E_i is the instantaneous modulus of elasticity of the concrete;

– n_{02} is the value of action n at equilibrium state 2 at the level considered (without taking into account the behavior coefficient);

– the values of the function G (ψ, u/H_t) were defined according to the diagrams in Figure 8.9.

Silo category	1	2	3
Value of μ	1/200	1/300	1/400

Table 8.25. *Values of coefficient μ*

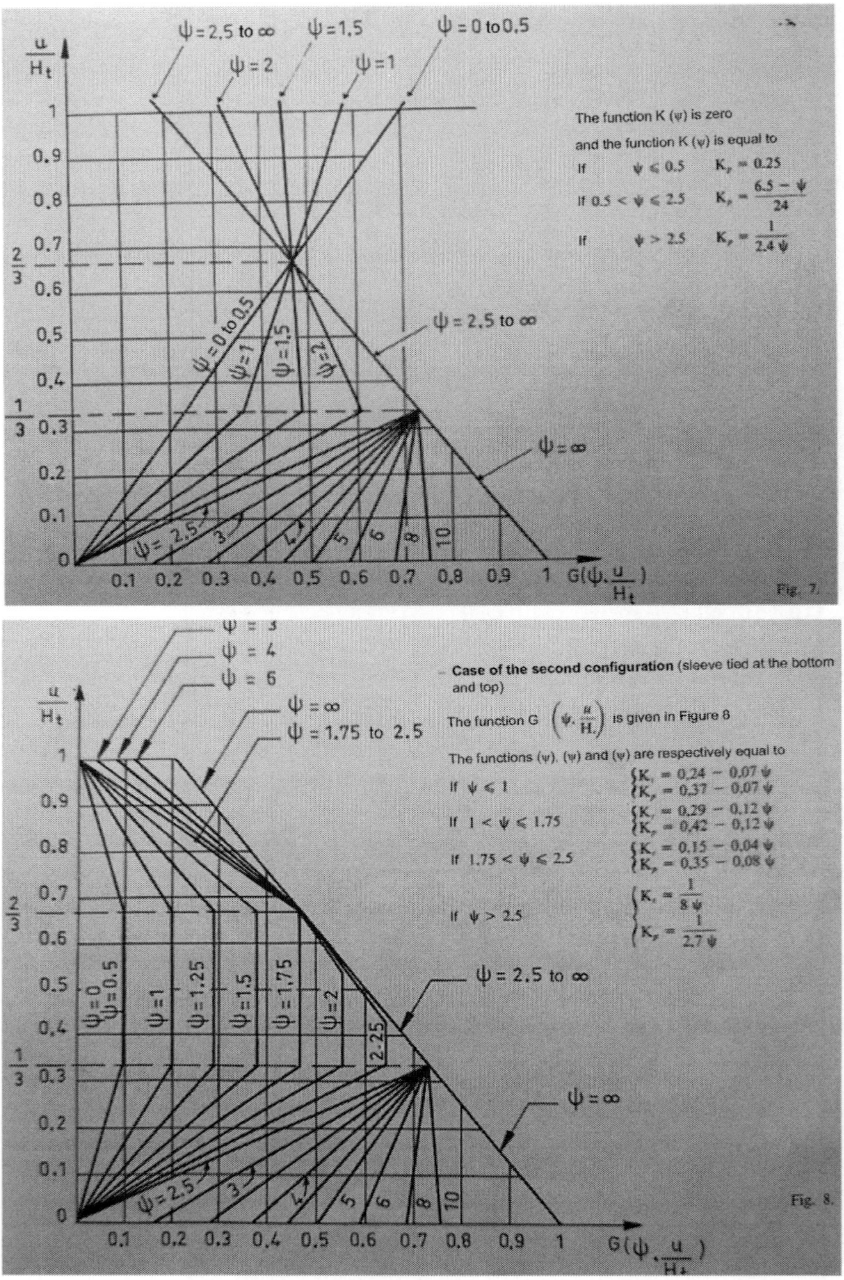

Figure 8.9. *Values of the function G (ψ, u/Ht)*

In the case of the third ovalization configuration, the following values were adopted:

$$G(\psi, u/H_t) = 1 - (4u/5H_t)$$

$$K_t(\psi) = K_p(\psi)$$

8.11.4.2. *Determining nominal actions on the vertical walls and the bottom of a silo with another type of abnormal emptying*

The professional rules also indicated provisions to be considered for air-blast silos and homogenization silos. It was common to fall back on the two states previously described.

8.11.5. *Special requirements for reinforced concrete structures*

The *actions* to be taken into account when determining the dimensions of structures according to CCBA 68 were as follows:

– permanent loads G (self-weight, fixed equipment, etc.);

– operating loads P (siled material, mobile equipment, etc.);

– climatic loads V (wind, snow);

– accidental SI loads (earthquake, explosion, etc.) when the CCTP provides for it;

– others mentioned in the available documents;

– deformations imposed by the temperature and thermal gradient T;

– the effects of shrinkage.

The combinations of the CCBA 68 regulations corresponded to the balanced load effects:

– first-type combination:

$$(S_1) = (G) + 1.2(P) + (T)$$

$$(S'_1) = (G) + (P) + (T) + (V)$$

– second-type combination:

$$(S_2) = (G) + 1.5(P) + (T) + 1.5(V)$$

$(S'_2) = (G) + (P) + (T) + 1.75(V)$

$(S''_2) = (G) + (P) + (T) + (SI)$

– service combination:

$(S) = (G) + (P) + (V) + (T)$

Stress	First type	Second type
Concrete (compression) $\overline{\sigma}'b$	– $\sigma'_b = 0.30\ (1 + 2e_r)\ \sigma'_{28}\ \psi_h$ if $e_r = M_v/(h_0N_v) \leq 0.5$ – $\overline{\sigma}'_b = 0.60\ \sigma'_{28}.\psi_h$ otherwise	$1.5\mu\,\overline{\sigma}'_b$
Steel (traction)	$2/3\ \sigma_{en}$	σ_{en}

Table 8.26. *Values of allowable stresses*

Allowable stresses with respect to the mechanical resistance were set as in Table 8.26.

Stresses associated with cracking were defined in the contract documents for each structure according to the following criteria:

– sensitivity of the ensiled product to moisture;

– residence time of stock in a cell;

– presence of an interior or exterior cover;

– requirement for air impermeability;

– situation in an aggressive atmosphere (seaside, etc.);

– relative position of the wall (inner, outer, tile, etc.).

These criteria made it possible to define three categories of walls:

– first category walls for which the probability of cracks appearing had to be as low as possible;

– second category walls for which crack opening was possible but whose value had to remain low;

– third category walls for which no stresses were imposed with regard to cracking.

The allowable tensile stress of steel was defined in CCBA 68 by:

$$\bar{\sigma}_2 = 1.32\ (\eta.k.\sigma_{28}/\phi)^{1/2}\ (MPa)$$

with:

– η: reinforcement cracking coefficient;

– Φ: bar diameter (mm);

– k (N/mm): parameter dependent on the wall category:

- k = 80 × 10³ N/mm for the first category,

- k = 100 × 10³ N/mm for the second category;

– σ_{28}: nominal tensile strength of concrete given by: $\sigma_{28} = (0.7 + 0.06\ \sigma'_{28}).\psi_b$ (MPa), with:

- $\psi_b = 1$ when the temperature of the concrete did not exceed 100°C,

- $\psi_b = 1 - (t - 100)/500$ if t was between 100 and 200°C.

NOTE.– A maximum tensioned reinforcement percentage was defined, which could lead to thickening of the walls. This maximum percentage was $\omega tf = A/(bh_0)$, where A represents the cross-section of the steel in a given direction over width b. It was limited to $\omega tf_{max} = 0.5/(\sigma_a/\sigma_{28} - n)$, where σ_a is the maximum stress of steels under service stresses and n = 15.

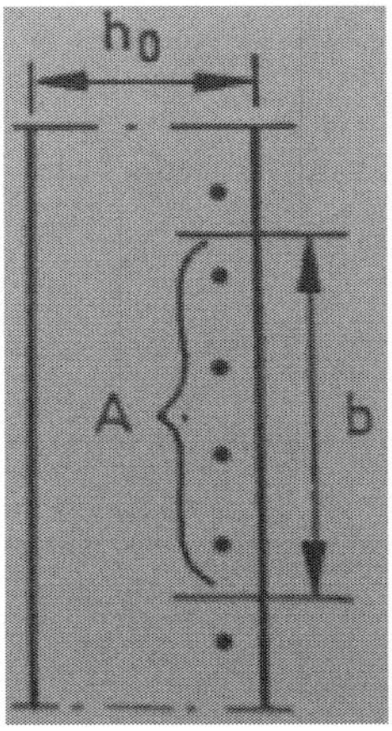

Figure 8.10. *Maximum reinforcement percentage*

8.11.6. *Example: determining the dimensions of a reinforced concrete silo*

Consider a silo composed of circular cells 41.50 m high and 9 m in diameter. The wall thickness is set at 0.18 m. These cells are connected to each other by an “ace of diamonds” in walls 0.22 m thick, its perimeter is 21.53 m and its surface area is 18 m^2.

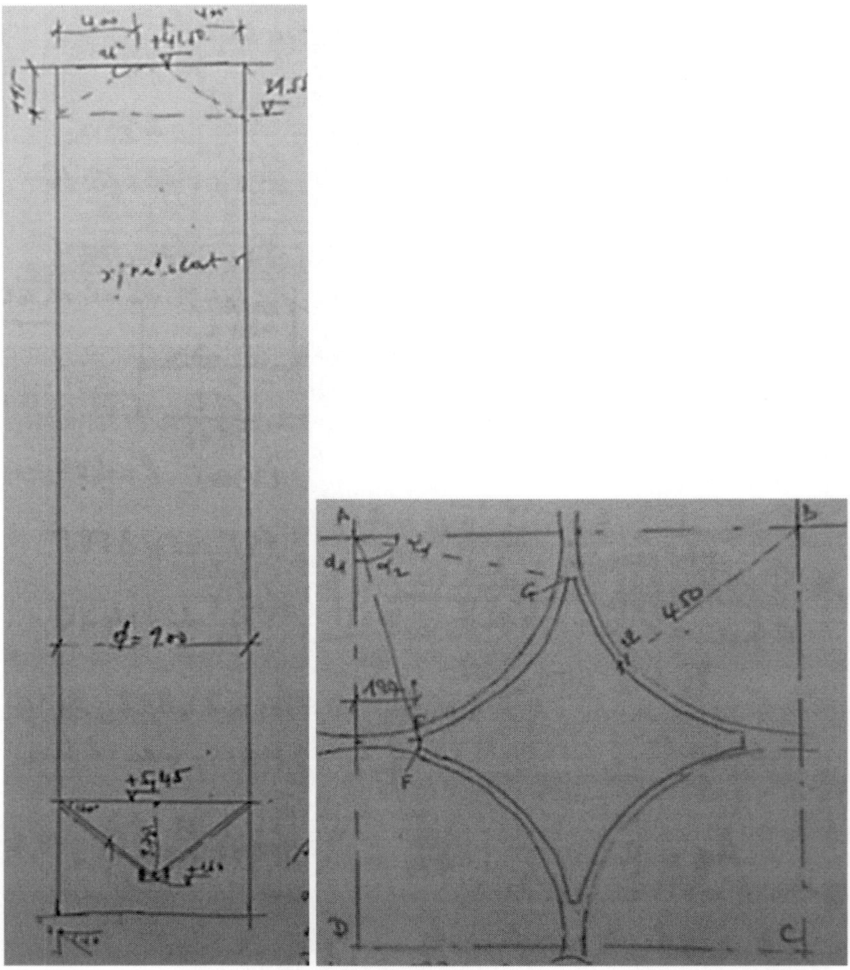

Figure 8.11. *Geometrical drawings of the cell and the ace of diamonds*

The assumptions taken into account for the sizing of the walls are as follows:

– $\gamma = 8.35$ kN/m^3 and $\varphi = 26°$ (wheat silos);

– $\delta = 20.09°$ (concrete wall type B);

– base plane +5.45;

– upper floor +41.50;

– reinforcement beam: 41.50–4.00 tg26° = +39.55;

– medium filling plane: 39.55 + 1.95/3 = +40.20, that is, a height of h = 40.20 – 5.45 = 34.75 m;

– normal emptying $r_H = 4.50/2 = 2.25$ m;

– $\rho = tg\delta/tg\varphi = 0.75$, $m = (1 - \rho^2)^{1/2} = 0.66$ and $h'' = 1/2\ r_H\ tg\delta = 0.41$ m.

The combined stresses on the cell walls are calculated according to the three states mentioned above:

– state 1: $n_1(z) = k_n\gamma\lambda_1 z_0\ (1 - \exp\ (-(z - h'')/z_0))$ with $\lambda_1 = (1 - m\ \sin\varphi)\ \cos^2\delta/(1 + m\ \sin\varphi) = 0.48$ and $z_0 = r_H/(\lambda_1\ tg\delta) = 2.25/(0.48.tg20.09) = 12.82$;

– state 2: $n_2(z) = k_n\ \gamma\lambda_2 z_0\ (1 - \exp\ (-(z - h'')/z_0))$ with $\lambda_2 = \cos^2\delta = 0.88$ and $z_0 = r_H/(\lambda_2\ tg\delta) = 7.00$.

The numerical values of the nominal loads are summarized in Table 8.27.

z	Height (m)	n (kPa) state 1	n (kPa) state 2
0	40.20		
3.46	36.74	12.60	20.90
6.92	33.28	23.70	35.81
10.38	29.82	32.12	44.90
13.84	26.36	38.60	50.44
17.30	22.90	43.45	53.80
20.76	19.44	47.20	55.85
24.22	15.98	50	57.10
27.68	12.52	52.15	57.84
31.14	9.06	53.80	58.33
34.60	5.60	55.10	58.61
34.75	5.45	55.12	58.62

Table 8.27. *Nominal load values*

It is also deduced that $\bar{v} = k_v\ \gamma\ (z_0.\bar{y} + h'') = 1.35 \times 8.35 \times (12.82 \times \bar{y} + 0.41) =$ 138 kPa for state 1 and $\bar{v}$ = 82.7 kPa for state 2 (the value of $\bar{y} = 1 - \exp(-x)$, where $\bar{x} = h/z_0 = 2.71$ for state 1 and 4.96 for state 2).

$T = \gamma.r_H.z_0.\ (x - y)$ (result of the friction forces on a section of unform width over the entire height of the silo above the base plane) is then worth T_1 = 423 kPa for state 1 and T_2 = 514 kPa for state 2.

It is possible to calculate in the same way the numerical values of the nominal effects at the ace of diamonds level with the following assumptions:

– γ = 8.35 kN/m^3 and φ = 26° (wheat silos);

– δ = 20.09° and r_H = 0.85 m.

We deduce the pressure values n according to the different equilibrium states (Table 8.28).

z	Height (m)	n (kPa) state 1	n (kPa) state 2
0	41.50		
4.76	36.74	13.81	18.47
8.22	33.28	18.22	21.34
11.68	29.82	20.36	22.11
15.14	26.36	21.41	22.32
18.60	22.90	21.92	22.38
22.06	19.44	22.16	22.39
25.52	15.98	22.28	22.40
28.98	11.52	22.34	22.40
32.44	9.06	22.37	22.40
35.90	5.60	22.38	22.40

Table 8.28. *Pressure values*

The values of $\bar{v}$ become: $\bar{v}_1$ = 55.89 kPa and $\bar{v}_2$ = 31.57 kPa.

The values of T are: T_1 = 220.4 kPa and T_2 = 235.8 kPa.

The cell walls are reinforced by considering the loaded cell alone and empty aces of diamonds. The base plane is set at +5.45, and the cell is loaded to +41.50.

Height	Traction in the wall T (kg)	Calculated steel cross-section (cm^2/lm) for a category 1 wall		Calculated steel cross-section (cm^2/lm) for a category 2 wall	
		Per CCBA	Per the regulations	Per CCBA	Per the regulations
Between +5.45 and +20	T = 58.62 × 4.50 = 26,400 kg/lm	18.6 cm^2	15 cm^2	13.2 cm^2	13 cm^2
Between +20 and +26.00	T = 53.80 × 4.50 = 24,200 kg/lm	17 cm^2	14 cm^2	12 cm^2	12.6 cm^2
Between +26.00 and +30.00	T = 50.44 × 4.50 = 22,700 kg/lm	16 cm^2	13 cm^2	11.5 cm^2	12 cm^2
Between +30.00 and +35.00	T = 44.90 × 4.50 = 20,200 kg/lm	14 cm^2	11.5 cm^2	10.1 cm^2	10.5 cm^2
Between +35.00 and +41.50	T = 20.90 × 4.50 = 9,405 kg/lm	6.5 cm^2	5.5 cm^2	5.0 cm^2	5.0 cm^2

Table 8.29. *Steel cross-section values*

The traction in a horizontal section of the wall is given by T = n.R, where n is the pressure calculated previously and R is the radius of the cell (see Table 8.29).

The allowable tensile stress of steel is given in the professional regulations by:

$$\bar{\sigma}_2 = 1.32\ (\eta.k.\sigma_{28}/\phi)^{1/2}\ \text{(MPa)}$$

with:

– $\eta = 1$ for smooth circular bars and $\eta = 1.6$ for high-grip reinforcements;

– $k = 80.10^3$ N/mm for category 1 and 100×10^3 N/mm for category 2;

– $\sigma_{28} = (0.7 + 0.06\ \sigma'_{28}).\psi_b = (0.7 + 0.06 \times 25).1 = 2.2$ MPa.

For a value of $\bar{\sigma}_2 = 175$ MPa in category 1 and 195 MPa in category 2.

The CCBA 68 set the reference tensile stress at 0.7 MPa for controlled concretes dosed at 350 kg/m^3. If we consider this last value, the allowable tensile stress for mild steel with a diameter of 10 mm becomes $\bar{\sigma}_2 = 142$ MPa for category 1 walls and $\bar{\sigma}_2 = 200$ MPa for category 2 walls.

NOTE.– The maximum percentage of wall reinforcement is $\omega tf_{max} = 0.5/(\sigma a/\sigma_{28} - n)$, that is:

– 0.40% for a category 1 wall;

– 2% for a category 2 and 3 wall.

In this example, the values would then be 7.20 cm^2/lm and 36 cm^2/lm. Therefore, the wall would have to be thickened for a category 1 cell. It would also be appropriate to take into account thermal effects in the following forms:

– gradient in wall thickness;

– gradient between opposite sides (sunny side/shaded side);

– expansion hindered by the ace of diamonds.

The ace of diamond walls are generally linked to the cell walls by columns. These columns are considered embedded at the level of the girdles (level +5.45). The walls are then subjected only to compressive forces due to the thrust of the ensiled material. Above this height, the columns are subjected to bending stresses, which also generate bending in the walls.

When calculating the stresses in the ace of the diamonds, two cases were considered:

– Case 1: only the circular cells are full, and the ace of diamonds is empty; then, only the ace of diamonds is full, and the cells are empty.

– Case 2: the circular cells surrounding the ace of diamonds are full at the same time as the latter, and the stresses are superimposed by applying a reducing coefficient of 0.85 to the stresses of the ace of diamonds alone.

9

BAEL 80 Regulations

9.1. Overview

As mentioned in the previous chapter concerning the CCBA 68 regulations, it had been envisaged as early as 1971, when the “common directives for the calculation of constructions” (DC 71) were published to move from a deterministic “allowable stresses” calculation method to a semiprobabilistic “limit states” method. However, engineers were not sufficiently prepared, and a transition period was desirable. New common directives were then published in 1979 (booklet 79–12 bis).

The “allowable stresses” calculation was based on the fact that determining the dimensions using “resistance of materials” rules made it possible to obtain the effective stresses that existed at a point of the structure under a given stress. An allowable stress was then set, which, when exceeded, could cause danger or harm.

The main disadvantages of this method (which has been applied since 1906) are as follows:

– Failure to take into account the possibility of cross-section plasticization (which permitted an increase in stress beyond the elastic limit).

– When the load effects are not proportional to the loads or the stresses are not proportional to the load effects, the safety coefficient applied to the stresses did not guarantee an increase in the loads. This was the case for parts subjected to composite bending, for example.

– Depending on the type of construction and the relative intensity of permanent and variable loads, the action of the latter can be very different, and it would be interesting to be able to differentiate, according to their nature, the effects of the different stresses, which was not possible with the allowable stress method.

– The reference to a single allowable stress did not allow the distinction to be made between the effect of directly applied loads and that resulting from imposed deformations.

The concept of limit state verification answered the question: what level of safety do we want to achieve? It was under the aegis of the Euro-International Concrete Committee (CEB) that this question was answered as follows.

The CCBA 68 regulations established the first type of verifications, which were supposed to cover the serviceability limit states, and those of the second type, which were supposed to cover the ultimate limit states. However, the second type of verifications had most often been neglected, so the overall safety of the constructions was not always verified. The principle of the semiprobabilistic rules then consisted of analyzing the factors involved in determining the safety of the structure and taking into account only the main elements through partial safety coefficients.

The BAEL regulations were validated by Circular No. 80–70 of May 23, 1980. These technical design and calculation regulations are applied to reinforced concrete constructions and structures using the limit states method. The first part of the instruction, entitled “General Rules,” gave the general principles and verification methods applicable to most constructions (chapters I to XIII of CCBA 68). The second part addresses the regulations applicable to standard reinforced concrete structures (chapter XIV of CCBA 68). It exclusively concerned buildings.

The coexistence of the BAEL and CCBA 68 regulations was the subject of Appendix II of the decree of October 16, 1979. The latter set the coexistence at two years. During this period, public works or building projects could be sized according to the CCBA 68 or the BAEL regulations but without mixing them.

The BAEL regulations did not apply:

– to unreinforced concrete structures;

– to lightweight aggregate concrete structures;

– to cavernous or reinforced cellular concrete structures;

– to elements reinforced with rolled profiles and mixed steel/concrete structures;

– to elements subjected to temperatures deviating from purely climatic influences (industry, atomic pile chambers, etc.).

The regulations were structured as follows:

I) General rules:

– Chapter A1: Principles and definitions;

– Chapter A2: Characteristics of materials;

– Chapter A3: Actions and stresses;

– Chapter A4: Justification of prismatic parts subjected to normal load effects;

– Chapter A5: Justification with regard to tangent load effects;

– Chapter A6: Grip;

– Chapter A7: Various construction provisions;

– Chapter A8: Special provisions for certain elements.

II) Regulations applicable to frames and common elements of reinforced concrete structures:

– Chapter B1: Indications on the choice of concrete according to the strengths to obtain and the manufacturing conditions;

– Chapter B2: Scope of application;

– Chapter B3: Load effect analysis;

– Chapter B4: Connections between the various construction elements;

– Chapter B5: Effects of size variations and differential settlements;

– Chapter B6: Floors and beams;

– Chapter B7: Slabs on continuous supports;

– Chapter B8: Columns;

– Chapter B9: Foundations.

The circular of May 23, 1980, also specified that the BAEL regulations could be applied to existing structures subject to the following adaptations:

– introduction of actual values of applied dead loads;

– concrete strength (characteristic strength);

– values of the applicable coefficients;

– frequent operating loads to consider.

The following notations were adopted:

– A or A_s: cross-sectional area of steel;

– A_t: sum of the cross-sectional areas of a transverse reinforcement;

– B: concrete cross-section;

– E_s: modulus of elasticity of steel;

– E_b: longitudinal deformation modulus of the concrete;

– E_{fl}: creep deformation modulus;

– E_i: instantaneous deformation modulus (E_{ij} at j days);

– E_V: delayed deformation modulus (E_{vj} for a load applied at j days);

– c_g: aggregate size;

– f_{cj}: characteristic compressive strength of the concrete at j days;

– f_{tj}: characteristic tensile strength of the concrete at j days;

– σ_{bc}: compressive stress of the concrete;

– σ_{st}: tensile stress of steel (or σ_s);

– τ: tangent stress;

– τ_s: adhesion stress;

– τ_{se}: traction drive stress;

– φ: creep coefficient;

– ψ_s: sealing coefficient of a reinforcement.

9.2. Calculation bases: requirements

9.2.1. *Basic requirements*

The calculations were carried out using the semiprobabilistic method according to the following limit states:

– the ultimate limit states for static equilibrium, strength and form stability limits;

– the serviceability limit states which targeted operating and durability conditions. These studies focused mainly on the deformation limit states (instantaneous and delayed) and crack opening.

NOTE.– The ultimate limit states did not include the fatigue limit state. When fatigue verification had to be carried out, the CCTP designated the parts of the structures or structures of interest and specified the calculation methods in terms of effects, combinations and coefficient values.

NOTE.– These regulations are only applied to reinforced concrete structures made from normal natural aggregates and contain a minimum cement content of 300 kg/m^3.

9.2.2. *Project service life*

The service life was not explicitly mentioned in the regulations. However, the provisions adopted (coating, concrete strength, etc.) assumed that the service life of buildings was at least 50 years.

9.2.3. *Thermal effects*

Chapter B5 specified the regulations to be adopted with respect to dimensional variations and differential settlements for standard reinforced concrete elements. The effects of these variations were generally neglected in ultimate limit state verifications and were only considered in serviceability limit state calculations.

For standard and industrial constructions, the effects of shrinkage and temperature variations were disregarded if there were expansion joints at a maximum distance of:

– 25 m in the neighboring Mediterranean regions;

– 30–35 m in the eastern regions of France, the Alps and the central mountain range;

– 40 m in and around Paris;

– 50 m in western France.

NOTE.– It should be noted, however, that for rectangular buildings with hard cores (stairwells or elevator shafts, stability platforms, etc.) at both ends and respecting the distances mentioned above between joints, cracks at 45° could appear at the floor level. Placing hard cores in the center of the building and reinforcing the floors of the extreme bays by implementing reinforcements parallel to the bisectors of the angles were recommended. Additionally, for slight excesses of the previous values, it was possible to take into account only a fraction α of the effects of shrinkage and expansion according to the following regulations:

– if $l \leq l_{max}$, then $\alpha = 0$;

– if $l_{max} \leq l \leq 1.25\ l_{max}$, then $\alpha = 4(l/l_{max} - 1)$;

– if $1.25\ l_{max} < l$, then $\alpha = 1$.

As the upper floors are more sensitive to thermal variations than the others, it was recommended to create tuning fork-shaped joints between the last floor and the terrace. On the other hand, the expansion joints were not necessarily extended into the buried parts and the foundations.

The effects of dimensional variations in the vertical direction of the structures had to be taken into account when the load-bearing elements were located outside the thermal barrier, for example, for high-rise buildings where the columns are placed outside the facades or when the thermal insulation is located in the building interior (effect of overall vertical expansion and effect of the thermal gradient). The length of the exposed elements was limited to:

– 6 m in humid and temperate regions;

– 4 m in dry regions or regions with strong thermal opposition.

The reinforced sections were at least equal to 0.2% of the concrete cross-section for concretes with a tensile strength greater than 2.4 MPa and 0.25% in other cases.

The previous steel cross-sections were doubled if the lengths between the expansion joints exceeded 12 m for humid regions and 8 m for dry regions.

9.2.4. *Differential settlements*

For dimensional variations due to thermal effects, it was accepted not to consider differential settlements for current structures as long as the foreseeable difference in level at the serviceability limit state remained lower than the following values:

– 1/300 of the span between interior and exterior load-bearing elements;

– 1/500 of this scope when only settlement effects are involved;

– 1 cm in the case of rigid or fragile partitions blocked in the frame;

– 2 cm in other cases.

9.2.5. *Material properties*

The properties of the materials used for reinforced concrete are given in Chapter A.2 of the BAEL regulations and repeated in section 9.3.

9.3. Materials

9.3.1. *Concrete*

9.3.1.1. *Strength*

The characteristic strength of concrete was defined by a 28-day strength value f_{c28} measured on straight cylinders with a cross-section of 200 cm^2 and a height equal to twice the diameter.

For stresses occurring within j days (before 28 days), the compressive strength of the concrete f_{cj} was defined by:

$$f_{cj} = 0.685\ f_{c28}\ \lg\ (j + 1)$$

For large values of j (greater than 28 days, 90 days, for example), the value of the compressive strength was estimated by:

$$f_{cj} = 1.10\ f_{c28}$$

The characteristic tensile strength of the concrete at j days was given by the following formula:

$$f_{tj} = 0.6 + 0.06\ f_{cj}$$

This formula was limited to concrete whose $f_{cj} \leq 40$ MPa.

The main concrete qualities used are summarized in Table 9.1.

Concrete quality	Cement dosage (kg/m^3)	Average strength at 28 days (MPa)	Characteristic compressive strength at 28 days f_{cj} (MPa)	Characteristic tensile strength at 28 days f_{tj} (MPa)
Low strength concrete	300	20–25	16	1.6
Standard concrete	350	25–30	20	1.8
High strength concrete	400	30–35	25	2.1
Ultra strong concrete	400 + admixture	35–40	30	2.4

Table 9.1. *Strength of standard concrete*

NOTE.– Compared with those of the CCBA 68 regulations, the characteristic values of the BAEL regulations are approximately 15% lower than the previous nominal values of the CCBA regulations, simply because of the way in which strength is defined.

The values of 16–20–25 and 30 MPa therefore correspond to the nominal values of 18–23–28 and 34 MPa of CCBA 68.

9.3.1.2. *Elastic deformation*

– *Instantaneous longitudinal deformation*

Under normal stresses of a *duration less than 24 h*, the instantaneous longitudinal deformation modulus of the concrete E_{ij} is:

$$E_{ij} = 12{,}000\ f_{cj}^{1/3}\ (\text{MPa})$$

– *Delayed longitudinal deformation*

To account for the effects of shrinkage and creep, the value of the longitudinal deformation modulus was modified to:

$$E_{vj} = 4{,}000\ f_{cj}^{1/3}\ (\text{MPa})$$

By default, it was assumed that the additional long-term deformations due to creep and shrinkage were equal to twice the short-term deformations.

The Poisson's ratio of the uncracked concrete was 0.2. It was taken as 0 for cracked concrete.

9.3.1.3. *Shrinkage*

The shrinkage coefficients reached the following values:

– 3×10^{-4} in southeastern France;

– 2×10^{-4} elsewhere.

9.3.2. *Reinforced concrete steels*

9.3.2.1. *Properties*

The BAEL regulations referred to booklet 4 of the CCTG. The mechanical and technological properties of steels were defined by:

– the elastic limit f_e (conventional limit at 0.2%);

– tensile breaking stress f_s;

– elongation at break A;

– the absence of defects;

– cold bending capacity;

– the absence of fragility;

– welding capacity.

Two characteristic diagrams were then proposed for the calculations:

– type 1 diagram (case of natural or highly work-hardened steels) (see Figure 9.1);

– type 2 diagram (case of work-hardened steels for which high precision was required) (see Figure 9.2).

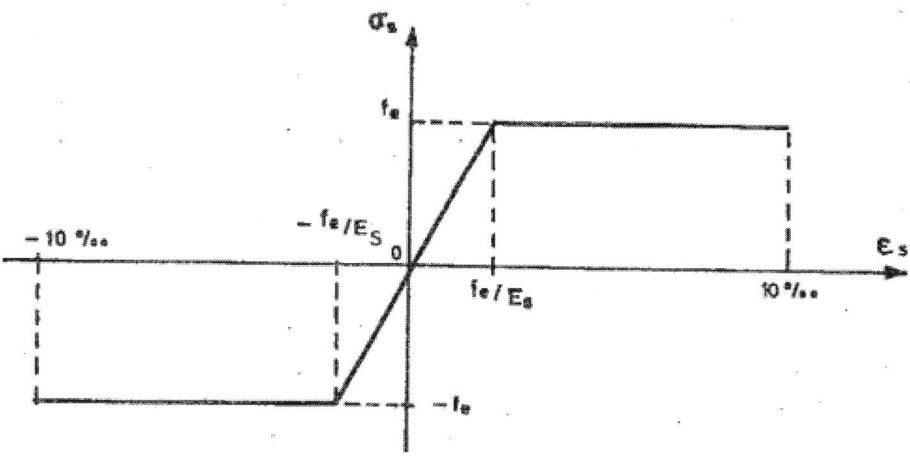

Figure 9.1. *Characteristic diagram of type 1 steel*

The modulus of elasticity of the steel was 200,000 MPa.

The different types of approved steel were as follows:

– smooth round steel (according to standard A35.015). These were mainly the natural steels FeE22 and FeE24;

– high-grip bars obtained by hot rolling a naturally hard steel (type 1 steel);

– high-grip bars obtained by hot rolling followed by work hardening without cross-section reduction (type 2 steel);

– high-grip bars obtained by hot rolling followed by work hardening by wire drawing or cold rolling, resulting in a significant reduction in the cross-section (type 3 steel).

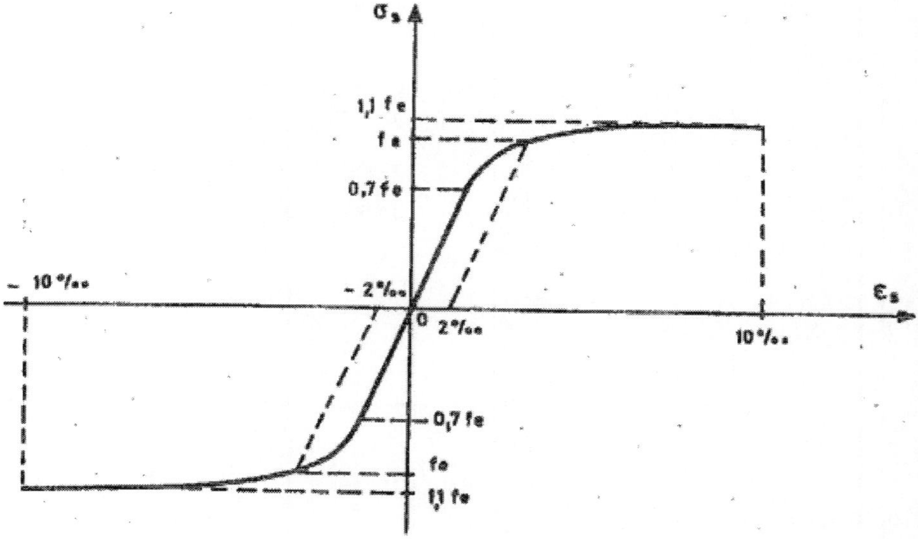

Figure 9.2. *Characteristic diagram of type 2 steel. For $\sigma \geq 0.7\ f_e$, the equation of the curve was given by: $\varepsilon_s = \sigma_s/Es + 0.823\ (\sigma_s/f_e - 0.7)^5$. For $\sigma \leq -0.7\ f_e$, the equation of the curve was given by: $\varepsilon_s = \sigma_s/Es + 0.823\ (\sigma_s/f_e + 0.7)^5$*

9.3.2.2. *Strength*

There were four main grades of reinforcing steel, as summarized in Table 9.2.

	Grade	Elastic limit f_e (MPa)	Breaking stress f_s (MPa)	Elongation at break (%)
Natural steel	FeE22	215	330–490	22
Smooth round	FeE24	235	410–490	25
High-grip steel	FeE40	400	480	14
	FeE50	500	550	12

Table 9.2. *Mechanical characteristics of common steel*

9.3.2.3. *Steel/concrete adhesion*

Chapter A6 of the rules specified the bonds between steel and concrete. For bars, these bonds were measured by the bond stress τ_s, whereby:

$$\tau_s = dF/(u.dx)$$

with:

– dF/dx: variation per unit length of the axial force exerted on the reinforcement;

– u: useful perimeter of the reinforcement (nominal perimeter for an isolated bar, for a bundle of two bars $u = (\pi + 2)\ \phi$; for a bundle of three bars $u = (\pi + 3)\ \phi$).

For bar anchors, the anchoring of an entire bundle of bars was not permitted; the bars had to be anchored individually. The adhesion stress on an anchor was assumed to be constant and equal to:

$$\tau_s = 0.6\ \psi s^2\ f_{tj}$$

with:

– $\psi_s = 1$ for smooth round bars;

– $\psi_s = 1.5$ for high grip bars.

9.4. Durability and cover of reinforcements

9.4.1. *Overview*

The durability requirements were not explicitly specified, as they are in the Eurocodes.

9.4.2. *Durability requirements*

The BAEL did not provide durability criteria. However, Chapter B1 specified how to choose concrete based on strength criteria, manufacturing conditions, risks of "deterioration" and risks related to external conditions.

– Thus, for the protection of "exposed" structures with a risk of corrosion of steel but without particular aggressiveness, a minimum dosage of CPA and CLK cement was recommended, given by the formula $550/(c_g)^{1/5}$. For common concrete, the dosage values are summarized in Table 9.3.

Granular class filter c_g	Minimum content
12.5	330
20	300
25	290

Table 9.3. *Minimum cement content in concrete for protection against steel corrosion*

– For structures exposed to harsh conditions, such as sea water, sea spray and selenite concentrations in water greater than 5,000 mg/L, the minimum cement content was given by the formula $700/(c_g)^{1/5}$. The values are given in Table 9.4.

Granular class filter c_g	Minimaum content
12.5	420
20	385
25	370

Table 9.4. *Minimum cement content in concrete in a marine environment*

NOTE.– When other types of cement (CPJ, etc.) are used, it is necessary to provide proof of equivalent active characteristics.

9.4.3. *Cover*

The covers defined as protecting the reinforcements (in the sense of the durability of the structures) were at least:

– 4 cm for structures at sea or exposed to sea spray or salt fog as well as for *structures subjected to very aggressive atmospheres*;

– 3 cm for unformed facings subjected to aggressive actions (e.g. the upper faces of bridge decks);

– 2 cm for facings exposed to bad weather, condensation and parts of structures in contact with liquids (tanks, pipes, etc.);

– 1 cm for the walls of structures located in covered and enclosed premises, not exposed to condensation.

9.5. Structural analysis

9.5.1. *Overview*

The purpose of the analysis is to determine the distribution of stresses. The model used by BAEL is that of linear elasticity calculation. Article A.3.2.1 thus specified that "generally speaking, stresses are calculated using an elastic and linear model for the structure. Material resistance methods are used to the extent that the shape of the parts allows it". However, this model could prove insufficient when calculating shape stability or accidental stresses.

It was then possible:

– to adopt patterns close to known structures;

– to use more elaborate models;

– to carry out experiments on a scaled-down model.

It was also accepted to adopt the following simplifications:

– For hyperstatic unknowns that do not depend on the modulus of elasticity, it was possible to calculate the elastic modulus in the uncracked section without considering the reinforcements.

– Second-order stresses were only considered for the ULS verification of form stability. For compressed parts with a slenderness less than 50, second-order stresses were taken into account on a flat-rate basis.

– For stresses dependent on Poisson's ratio, the values of the latter were taken equal to those mentioned in section 9.3.1.2.

– For parts embedded on supports and whose embedding conditions did not correspond to the elastic and linear model (e.g. partial blocking of the support rotation), it was accepted to take into account only a fraction of the maximum moment.

– For the ultimate limit state verification, it was not necessary to take into account the imposed deformations to the extent that the ductility or flexibility of the structures allowed it (e.g. not taking into account thermal effects if expansion joints are created).

9.5.2. *Load combinations*

Article A.3.3 recommended taking into account the following loads:

– G_{max}: all adverse dead loads;

– G_{min}: all favorable dead loads;

– Q_1: basic variable loads;

– Q_i: variable accompanying loads.

NOTE.– With respect to buildings, variable loads were set out in standard NF P 06.001 of April 1978. For road bridges, reference should be made to circular no. 71–155 of December 29, 1971. Climatic loads were subject to NV 65 regulations (revised in 1967 and 1978). Appendix D of the BAEL also provides the values of the normal snow loads (Sn). With respect to accidental loads such as earthquakes, shocks, explosions, fires, etc., the loads to be taken into account were specified in the CCTP.

9.5.2.1. *Ultimate limit state fundamental combination*

Nature of the base load	Nature of the accompanying load	Value of γ_{Qi}
Loads applied during execution	Wind	1.3
Loads on building floors	Wind or snow	1.2
Wind or snow loads on buildings	Loads on floors	1.04–1.3*

*The value depended on the building use; it was 1.04 for accommodations, classrooms, restaurants, offices, shops, collective kitchens, library reading rooms and large sales areas; it was 1.3 for storage facilities, archives, performance halls, industrial buildings and stands.

Table 9.5. *Values of coefficient γ_{Qi}*

The design stresses for the ultimate limit states of resistance and stability were the subject of the following *fundamental combination*:

$$1.35\ G_{max} + G_{min} + \gamma_{Q1}\ Q_1 + \Sigma\ \gamma_{Qi}\ Q_i$$

with:

– γ_{Q1} = 1.6 for road loads; 1.35 for narrowly defined or special loads (exceptional convoy, etc.); 1.5 in other cases;

– γ_{Qi} was defined according to the nature of the basic variable load (see Table 9.5).

9.5.2.2. *Ultimate limit state accidental combination*

The design stresses with respect to the ultimate limit states in accident situations were as follows:

$$G_{max} + G_{min} + F_A + \Sigma\ \gamma_{Ai}\ Q_i$$

where F_A is the representative value of the accidental load and γ_{Ai} represents the coefficient affecting the accompanying variable loads, that is, 0.5 for temperature. For the other loads, the values were fixed by the CCTP or other texts (e.g. paraseismic rules).

9.5.2.3. *Serviceability state limit combinations*

The general formulation of the combination of serviceability limit state loads was given by:

$$G_{max} + G_{min} + Q_1 + \Sigma\ \psi_{0i}\ Q_i$$

For bridges, road load Q_1 is multiplied by 1.20. The values of the coefficient ψ_{0i} are:

– 0.6 for thermal effects;

– in classic building cases (see Table 9.6).

Nature of the base load	Nature of the accompanying load	Value of ψ_{0i}
Loads applied during execution	Wind	1.0
Loads on building floors	Wind or snow	0.9
Wind or snow load on buildings	Loads on floors	0.8–1.0

Table 9.6. *Values of the coefficient ψ_{0i}*

9.5.3. *Second-order effects*

Article A4.4 of the regulations specified the justifications under second-order effects. The requests were calculated from the combinations of previous loads, taking into account:

– of an initial geometric imperfection consisting of:

- an overall inclination of 1/100 of a radian of the structure in the case of a frame,

- an additional eccentricity of the loads equal to the maximum (2 cm; l/250) (l being the length of the element) in the case of an isolated structural element;

– second-order stresses related to structural deformation. In common cases, the deformations were evaluated from the parabolic stress–deformation diagram and a creep coefficient of 2;

– section or mean line defects. The values were then fixed in the CCTP.

9.5.4. *Modeling the structure*

The digital modeling of structures was not considered in the BAEL regulations.

9.6. Structural calculations

9.6.1. *Simple compression*

The general requirements were defined in the first part of the regulations, in particular:

– section A4.3.5: case of composite bending with compression;

– section A4.4: limit states of form stability;

– section A6.1.24: compression bar junction;

– section A8.1: compressed elements.

Section B.8 was applied to the columns of current constructions. The combination of loads to be considered was given by:

$$1.35\ G + 1.5\ Q_B$$

The justification of the columns then consisted of determining the value of the resistant normal force, considering that the implementation conditions and, in particular, the flatness tolerances of the formwork were such that the straightness imperfection was less than the maximum (1 cm; l/500). In this case, the ultimate resistance normal force was given by the following formula:

$$N_u = \alpha\ [(B_r\ f_{c28})/(0.9\ \gamma'_b) + A\ f_e/\gamma'_s]$$

with:

– $\alpha = 0.85/(1 + 0.2\ (\lambda/35)^2)$ if $\lambda \leq 50$;

– $\alpha = 0.60/(50/\lambda)$ if $50 < \lambda \leq 100$ (the values of α are divided by 1.10 if more than half of the loads were applied before 90 days);

– A: cross-section of compressed steel considered in the calculation;

– B_r: reduced cross-section of the post equal to the actual cross-section minus 1 cm around the entire periphery;

– $\gamma'_b = 1.5$;

– $\gamma'_s = 1.15$.

λ is the slenderness of the column (the ratio of the buckling length to the gyration radius). The buckling length l_f was determined as a function of the free length l_0 and support conditions (see Table 9.7).

Assumptions	**Free at one end of the part and embedded at the other**	**Part hinged at both ends**	**Embedded part at both ends with possible horizontal movement**	**Part embedded at one end and hinged at the other**	**Part embedded at both ends without the possibility of movement**
Value of l_c	2 l_0	l_0	l_0	$l_0/\sqrt{2}$	$l_0/2$

Table 9.7. *Value of buckling lengths*

NOTE.– Notably, if the embedments are not perfect in terms of the RDM, the buckling lengths could be underestimated. In the case of standard buildings, which were braced by infills or stability beams and in which the continuity of the columns could be ensured, the buckling length was taken to be equal to:

– 0.7 l_0 if the column is embedded at its ends into the foundation or assembled into floor beams with the same rigidity;

– l_0 in other cases.

9.6.2. *Compression with flexion*

This case is described in section A4.3.5 of the BAEL regulations. The calculation of the stresses was carried out per ultimate limit states. The moment was established in relation to the center of gravity G of the cross-section with the incidence of the moment provided by the normal force N_u eccentric of e, or:

$$M_{Gu} = M_u + N_u \times e$$

The stresses at one moment M_{Gu} and normal effort N_{Gu} were applied to the center of gravity, and the moment M_{Au} calculated in relation to the tensioned steels was then:

$$M_{Au} = M_{Gu} + N_{Gu} \times (d - h/2)$$

where d is the useful height and h is the total cross-sectional height.

9.6.3. *Bending*

9.6.3.1. *ULS bending*

The ULS calculation was carried out based on the following assumptions:

– the plane sections are preserved;

– the tensile strength of the concrete is not considered;

– there is no slippage between the reinforcements and the concrete;

– the ultimate shortening of the concrete is: $\varepsilon_{bu} = 3.5 \times 10^{-3}$ during bending and 2.0×10^{-3} in centric compression;

– the elongation of steels is limited to 10×10^{-3};

– the deformation diagrams pass through one of three pivots, A–C, as specified in Figure 9.3.

In domain 1, the diagram passes through point A, corresponding to an elongation of approximately 10‰.

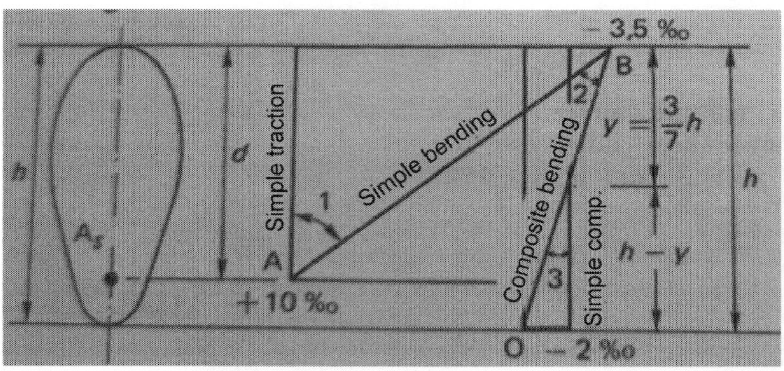

Figure 9.3. *Three pivot diagram*

In domain 2, the diagram passes through point B, corresponding to a concrete shortening of 3.5‰.

In domain 3, the diagram passes through point C, corresponding to a concrete shortening of 2‰ at a distance from the most compressed fiber equal to 3/7 of the total height h of the cross-section.

If $\alpha = y/d$, (y is the depth of the neutral axis from the upper fiber, and d is the useful height), the different domains then correspond to:

– domain 1 (pivot A): $y \leq 3.5\ d/(3.5 + 10) = 0.259\ d$, that is, $\alpha \leq 0.259$;

– domain 2 (pivot B): $0.259\ d \leq y \leq h$, that is, $0.259 \leq \alpha \leq h/d$;

– domain 3 (pivot C): $y \geq h$, that is, $\alpha \geq h/d$ (= 1.10).

The ULS calculation under bending stresses uses the concrete stress–deformation diagram, either according to the parabola-rectangle model or according to the simplified rectangular model, as shown in Figure 9.4.

When the parabola-rectangle diagram is used, the resultant compression force F_b for a rectangular cross-section of width b_0 is:

$$F_b = 0.81 f_{bu} \times b_0 \times y$$

The center of application of the force was located at a distance of 0.416y from the upper fiber, that is, a value of $z = d - 0.416y$.

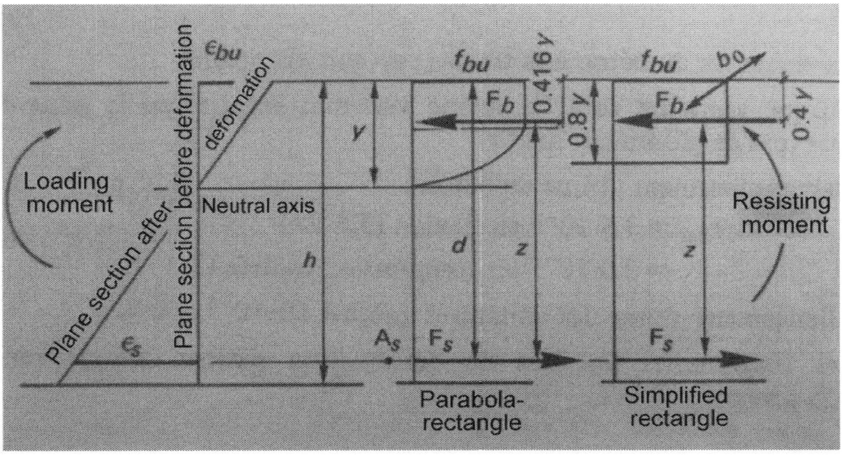

Figure 9.4. *Concrete stress–deformation*

In the case of the simplified rectangular diagram, $F_b = 0.8f_{bu} \times b_0 \times y$ with $z = d - 0.4y$.

The stress–deformation diagram of the steel was as described in section 9.3.2.1. The calculation methodology was based on the balance of compressive forces of the concrete, on the one hand, and tensile forces of the steel, on the other hand. It was assumed that the concrete had reached its maximum shortening (pivot B) and that its ultimate limit stress $f_{bu} = 0.85\ f_{cj}/\gamma_b$. On the other hand, the steel had reached its design elastic limit fe/γ_s, corresponding to a lengthening of ε_s.

The balance of forces implied, in simple bending, that $F_b = F_s$, that is, $0.8y\ b_0\ f_{bu} = A_s\ f_e/1.15$. Solving this equation allowed us to find y, the position of the neutral axis. The elongation of the steel was given by:

$$\varepsilon_s = \varepsilon_{bu}\ (d - y)/y$$

Several cases could then arise:

– $\varepsilon_s \leq 10$ ‰ (pivot B): from the stress–deformation diagram of steel, the value of the tensile stress was found; if this value was different from fe/1.15, it was calculated by iteration for y.

– $\varepsilon_s > 10$ ‰ (pivot A): the steel cross-section was insufficient, we recalculated $y = \varepsilon_{bu}/\varepsilon_s \times (d - y)$, that is, $y = 0.26d$, and the steel cross-section A_s was increased according to the formula: $A_s = 0.8y\ b_0\ f_{bu}/f_{su}$.

Typically, $f_{su} = f_e/1.15$ (for high-grip type 2 steel, this value was increased to $1.10\ f_e/1.15$).

9.6.3.2. *ULS bending*

The ULS bending calculation is generally used to determine the dimensions of structures in the case of harmful or very harmful cracking. The limits imposed for the stresses were such that the materials remained in the elastic domain, and the calculation of the latter was then carried out according to the classic RDM formulas, that is,

– maximum concrete compression: $\sigma_b = My/I$;

– maximum traction in the reinforcements: $\sigma_s = n\ M\ (y - d)/I$;

– maximum compression in compressed steel: $\sigma_s = n\ M\ (y - d')/I$.

With n, steel–concrete equivalence coefficient, or $n = E_s/E_b$, it was therefore possible to define a short-term and long-term value of n depending on the values of E_i and E_v. We then found $n = 7$ for the case of instantaneous loads and $n = 20$ for long-term loads. The BAEL retained the value of n = 15, corresponding approximately to a long-term module for dead loads and a semi-instantaneous module for variable loads.

The position of the neutral axis was given by the static moment equation:

$$b_0\, y \times y/2 + n\, A'_s\,(y - d') - n\, A_s\,(d - y) = 0$$

$$by^2/2 + n\,(A'_s + A_s)\, y - n\,(A_s\, d + A'_s\, d') = 0$$

The value of y was the positive root of this quadratic equation. Knowing y, it was then possible to calculate the moment of inertia I of the homogenized reduced cross-section, that is:

$$I = b_0 y^3/3 + n\, A'_s\,(y - d')^2 + n\, A_s\,(d - y)^2$$

Knowledge of I allowed the calculation of the ULS constraints and their comparison with the SLS defined in articles A.4.5.2 and A.4.5.3, that is:

- 0.6 f_{cj} for the compressive stress of the concrete;
- Min (2/3fe; 150η) MPa for the traction of steel with harmful cracking;
- Min (0.5fe; 110η) MPa for the traction of steel with very harmful cracking.

The total deformation had to be calculated by integrating the successive construction phases. Deformation due to bending was obtained by integrating the curvatures and those of the shear forces by integrating the distortions. The curvatures could be calculated in cracked or uncracked cross-section. In the second case, the cross-section was made homogeneous, with n = 15.

In the first case, the curvature was in the following form:

$$1/r = (\varepsilon_s + \varepsilon_{bc})/d$$

where 1/r denotes the deformation curvature and d denotes the useful height of the cross-section. ε_{bc} is the relative shortening of the compressed extreme-fiber concrete, and ε_s is the relative elongation of the steel under tension. These two values consider the deferred deformations of the concrete.

The distortions in the uncracked state were calculated in accordance with RDM formulas, considering the concrete alone. In the cracked state, a lattice analog was used.

9.6.4. *Buckling*

For more details, please refer to section 9.6.1.

9.6.5. *Justifications for tangent stresses*

These justifications mainly concerned beams subjected to shear forces (with the exception of deep beams and short consoles, which were the subject of Appendices F.5 and F.6). They were only performed for the ULS. They aimed to determine the transverse reinforcement of the core subjected to a shear stress τu such that $\tau_u = V_u/(b_0 \times d)$, where b_0 is the width of the core and d is the useful height.

For compressed parts, the requirements were only applied if $\tau_u \geq$ Max (0.04 f_{cj}; 1.5 MPa).

The justification for a standard cross-section depended on the level of cracking required for the structure. In the case of straight transverse reinforcements, the shear stress had to be such that:

– in minimally damaging cracking: $\tau_u \leq$ Min (0.13 f_{cj}; 4 MPa);

– in damaging and very damaging cracking: $\tau_u \leq$ Min (0.10 f_{cj}; 3 MPa).

For the transverse reinforcements inclined at 45°, these values increased to 0.18 f_{cj} and 5.5 MPa.

The minimum cross-section of the core reinforcements A_t must be such that:

$$A_t \times fe/(b_0\ s_t \sin\alpha) \geq \text{Max}\ (\tau_u/2;\ 0.4\ \text{MPa})$$

where b_0 is the core width, s_t is the spacing between two frames and α is the inclination of the frames.

Furthermore, the calculation of the core reinforcements was intended to balance the shear force obtained from the formula in section A.5.1.232:

$$\frac{A_t}{b_o s_t} \geq \frac{\gamma_s\left(\tau_u - 0.3 f_{tj}\ k\right)}{0.9 fe\left(\cos\alpha + \sin\alpha\right)}$$

By upper bounding f_{tj} to 3.3 MPa and with the value of k:

1 (unit) in simple bending.

$1+\dfrac{3\sigma_{cm}}{f_{cj}}$ In compound bending with compression, σ_{cm} denoting the average compressive stress of the total concrete section.

$1-\dfrac{10\sigma_{tm}}{f_{cj}}$ In compound bending with tension, σtm denoting the average tensile stress of the total concrete section, under the normal design force.

NOTE.– In the event of resuming concrete or very damaging cracking, k = 0.

For rectangular beams, the end connecting rod condition of the reinforcements made it possible to calculate the value of a (support length of the connecting rod) according to the following equation:

$$2\ V_u/(b_0\ a) \leq 0.8\ f_{cj}/\gamma_b$$

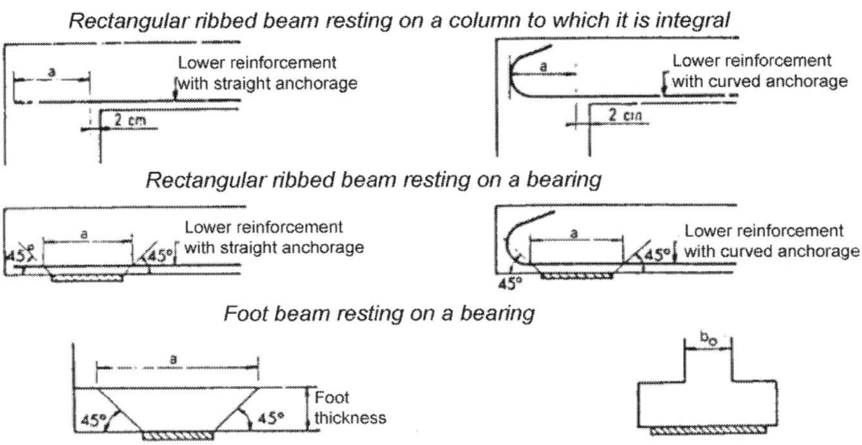

Figure 9.5. *Values to determine the value of a*

For the slabs and slab beams, no shear reinforcement was required if the following conditions were met:

– no resumption of concrete at the part height;

– $\tau_u \leq 0.05\ f_{cj}$.

9.7. Constructive provisions

The constructive provisions were included in chapter A.7 of the BAEL 83 regulations under the title “Miscellaneous constructive provisions”.

9.7.1. *Pouring concrete*

Section A.7.1 confirmed that to effectively protect steel against corrosion, it was necessary to have concrete with a sufficient cement content (350–400 kg/m^3) that was as impermeable as possible. This last condition was directly linked to the compactness of the concrete.

When needle vibration on concrete was planned, intervals between reinforcements were required for the passage of the needles (vibration chimneys). For tall parts, the same arrangement was used for casting (casting chimneys).

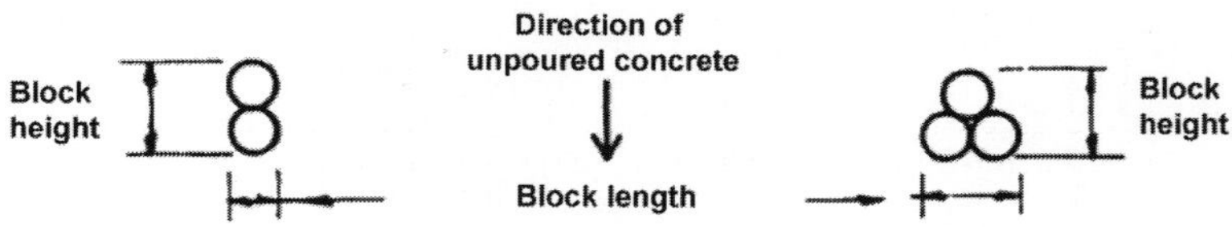

Figure 9.6. *Constructive arrangements of the reinforcements*

9.7.2. *Distance between reinforcements and between reinforcements and formwork*

Correct concreting was achieved by:

– a bar diameter used as reinforcement for slabs or walls less than or equal to 1/10 of the total thickness of the elements;

– a beam core reinforcement diameter less than Min (h/35; ϕl; 1/10 b_w) (h: total beam height; b_w: core width);

– grouping of bars into a bundle of two superimposed bars or three staggered bars. Groupings of more than three bars were permitted if they were not subject to any drive stress;

– a bar cover equal to the diameter of the bar if it is isolated to the width of the bundle otherwise;

– a distance between neighboring reinforcements equal to their diameter if they are isolated; otherwise, it is equal to the width of the bundle. This distance must be greater than 1.5 C_g horizontally (C_g is the diameter of the largest aggregate).

9.7.3. *Concrete recovery*

The concrete recovery surfaces had to meet the provisions of Article A.5.3, and if the recovery surface was not compressed, it must be rough.

9.7.4. *Vacuum thrust*

Article A.7.4 particularly targeted 45° bent frames that could generate vacuum thrusts (Figure 9.7).

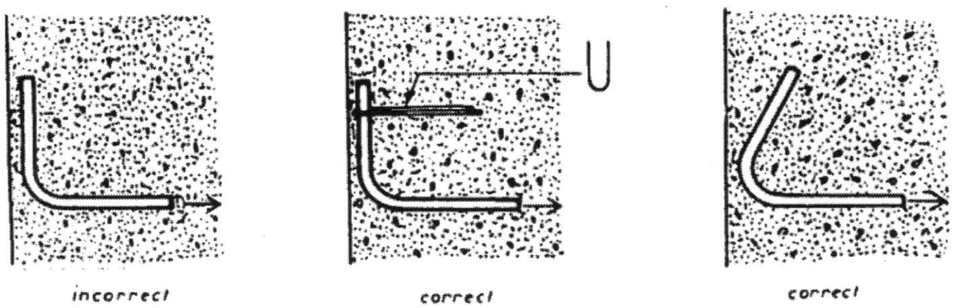

Figure 9.7. *Vacuum thrusts*

9.7.5. *Special provisions of compressed elements*

The provisions relating to the longitudinal reinforcement of the compressed elements were as follows:

– their cross-section is at least 4 cm²/mL of facing;

– their cross-section is at least equal to 0.2% of the concrete cross-section without exceeding the overlap zones by 5%;

– the maximum distance between two neighboring reinforcements on the same face must not exceed a + 10 cm or 40 cm (a being the length of the short side).

With respect to transverse reinforcements, the provisions mentioned in section A.8.1.3 are as follows:

– all longitudinal reinforcements with a diameter greater than 20 mm must be surrounded by a continuous transverse reinforcement;

– the diameter of the transverse reinforcements is at least equal to one-third of that of the transverse reinforcements;

– the transverse steel is less than the Min ($15\phi_l$; 40 cm; a + 10 cm) (ϕ_l is the diameter of the longitudinal reinforcements and a smallest dimension of the part);

– in the overlap areas of longitudinal reinforcements (one out of two), the number of rebar splices on the overlap is at least three.

9.7.6. *Slabs on continuous supports*

For slabs with a thickness between 12 cm and 30 cm, section A.8.2.4 provided the following results:

– The reinforcement ratio in each direction must be greater than 0.0012 for smooth bars, 0.0008 for high-grip FeE40 steels or welded wires with smooth wires with a diameter greater than 6 mm and 0.0006 for high-grip FeE50 steel or welded wires with smooth wires with a diameter less than 6 mm.

– The spacing of the steel in the same layer must respect the values in Table 9.8.

– The bending steel in the central zone of the slab must be extended on supports at the rate of their totality if the slab is subjected to mobile concentrated loads, otherwise at the rate of one in two steels.

Direction	**Only distributed loads**	**Concentrated loads**
Most stressed direction	3 h and 33 cm	2 h and 25 cm
Direction perpendicular to the most stressed	4 h and 45 cm	3 h and 33 cm

Table 9.8. *Spacing of slab steel*

9.8. Special case of tanks: booklet 74 of 1983

The main provisions adopted by booklet 74 in the 1983 version (decree 83–251 of March 29, 1983) were as follows.

9.8.1. *Classification of structures in terms of waterproofing*

Affected by the 1983 version are:

– reinforced concrete and prestressed concrete tanks;

– masonry tanks;

– ancillary works (valve chambers, metering, pumping and treatment station buildings, transformer stations, stores, etc.).

The classification in terms of waterproofing was defined as follows:

– When waterproofing the tanks is ensured by the building structure, it must be calculated in such a way as to limit cracks in accordance with the CCBA 68 regulations.

– When the tank structure has an added cover, the requirements to be respected are those of DTU 26.1 and booklet 64 for covers and the manufacturers' specifications for other processes. The degree of cracking of the support was not mentioned.

The content constraints are summarized in Table 9.9.

Type of concrete	Content (kg/m^3)	Minimum binder class
Blinding concrete	100	35
Mass-fill concrete for slope shapes, ballast, etc.	150	35
Unreinforced foundation and well concrete	200	35 or 45
Reinforced concrete not in contact with water	300	45
Reinforced concrete in contact with water	350	45
Prestressed concrete	400	See list in section 4 of booklet 74

Table 9.9. *Concrete composition*

NOTE.– FDP 18.011, which sets out the exposure classes for concrete, was published in 1992 and revised in 2009 and 2016. Before this date, care must be taken to ensure that the composition of the concrete matches the contents of the tank. Article 4 of the booklet specified that this class criterion might not be sufficient to take into account the aggressiveness of the environment but may not provide any additional information.

The dimensions of the aggregates were as follows:

– sand $0.08 \leq d \leq 6.3$ mm;

– grit $6.3 \leq d \leq 20$ mm;

– gravel $20 \leq d \leq 80$ mm.

NOTE.– In the current regulations, the maximum diameter of gravel has been reduced to 32 mm.

9.8.2. *Constructive provisions*

For tanks, the construction provisions related to the wall thickness were as follows:

– 0.10 m in the case of ordinary formwork if the wall only plays a mechanical role or supports a flexible waterproofing system;

– 0.12 m for structures that, in addition to the mechanical role, ensure waterproofing or support rigid waterproofing;

– 0.13 m and 0.16 m, respectively, when slip formwork is used.

For towers, the previous construction provisions became:

– 0.10 m in the case of ordinary formwork;

– 0.14 m for structures made using slip formwork.

9.8.3. *Loads*

The loads to be taken into account when determining the structural dimensions are those indicated in Table 9.10.

Note on the justification of tanks with CCBA 68:

– To justify the maximum opening of the cracks in the tanks, the booklet retained the formula for the tensile stress of the steel:

$\sigma_s \leq 2.4\ (k\eta\ \sigma_{tb}/\Phi)^{1/2} + 300\ \eta$ in bars

Loads	Nature	Value
Dead loads	– Dead weight of civil engineering – Dead weight of equipment – Weight of superstructures Prestressing	– According to NFP 06.001 – According to technical specifications – According to NFP 06.001 According to CCBA 68 or BPEL 80
Variable loads	– Weight and pressure of the contained liquid – Operating loads – Climatic loads – Maintenance loads – Loads on the ground surface – Thermal gradient – Earth weight and pressure – Water table underpressure Water pressure on the buried walls	– According to NFP 06.001 – According to NV 65 – 2 kN/m² for floors and walkways – 1 kN/m² for terraces To be specified in the specific requirements T_{max} and T_{min} in the specific requirements

Table 9.10. *Calculation of load values*

The values of k depend on the presence or absence of an inner coating. The formula in the 1998 version is used.

– The tensile stress of the concrete was also set to the value in the 1998 version.

– The opening of the cracks could therefore be considered limited to the following values (combination g + q):

- 0.10 mm for “class A” structures;

- between 0.10 and 0.15 mm for “class B” structures;

- greater than 0.15 mm for “class C” structures.

Note on the justification of tanks with BAEL 80:

We essentially find the same formulas as those for the 1998 version for the tensile stress of steel up to the following grade:

$\sigma_s \leq \alpha(\eta\ \sigma_{tb}/\Phi)^{1/2}$ in MPa

with:

– $\alpha = 240$ for self-watertight structures;

– $\alpha = 245$ for structures with a rigid mortar cover (adherent);

– $\alpha = 320$ for structures with an unreinforced flexible cover;

– $\alpha = 400$ for structures with a reinforced flexible cover;

– $\alpha = 450$ for structures with a semi-adherent or independent elastoplastic coating.

Thermal insulation:

Article 44 specifies the following elements:

– thermal insulation (in particular, that of the tank) must be indicated in the CCTP (no regulatory obligation);

– if it is planned, it must not be made of brick (or masonry in general).

9.9. Special case of tanks: booklet 74 of 1998

In the 1998 version (decree 98–39 of March 25, 1998), the previous provisions were modified as follows.

9.9.1. *Classification of structures in terms of waterproofing*

Affected by the version of 1998 are:

– water tanks (water towers, water tanks for various uses and fire-fighting reserves);

– sewage treatment plants;

– intake and drainage structures;

– channels (other than navigation);

– aqueducts;

– pumping and lifting stations, tanks and buffer basins;

– balance chimneys;

– storm basins and overflows;

– swimming pools;

– wine vats;

– slurry pits.

The watertightness classes of the structures were then defined as follows:

– *Class A construction*

DEFINITION 9.1.– Structure whose watertightness is ensured by the structure itself. In addition, the following constraints are imposed:

– $f_{c28} \geq 25$ MPa (compressive strength of concrete;

– $f_{t28} \geq 2.1$ MPa (tensile strength of the concrete);

– C (cement content) > 350 kg/m^3;

– reinforced concrete or prestressed concrete structure with possible incorporation of surface or body water repellent.

– *Class B construction*

DEFINITION 9.2.– Structure whose waterproofing is ensured by the structure and completed by a waterproofing coating (adherent interior screen but not resistant to cracking of the latter: hydraulic mortars, water repellents and unreinforced synthetic resins).

– *Class C construction*

DEFINITION 9.3.– Structure whose waterproofing is ensured by a waterproof coating (the structures have only a mechanical role).

EXAMPLE.– A plastic, elastoplastic coating was applied inside the structure and supported slight deformation and cracks in the support (membranes, reinforced resins).

– *Class D construction*

DEFINITION 9.4.– Structure constructed using prefabricated elements (the preceding provisions are applicable, including for the treatment of construction joints).

The composition of concrete and exposure classes were defined in the experimental standards XPP 18.305 and then NF EN 206/CN and in the booklet FDP 18.011, allowing ambient aggressiveness to be taken into account.

9.9.2. *Constructive provisions*

The minimum wall thickness was increased to:

– 0.15 m in class A;

– 0.12 m in classes B and C;

– 0.15 m for structures constructed using slipforming.

The minimum provisions for the reinforcement of the walls were as follows:

– two layers if the wall thickness is greater than 0.15 m;

– otherwise, one layer;

– for circular structures with a vertical axis:

- the inner layer does not comprise more than half of all horizontal reinforcements,

- the diameter of the reinforcement is greater than 8 mm and less than the wall thickness divided by 10;

– for walls in contact with liquid, the reinforcements are spaced less than 0.20 m and 1.5 times the thickness;

– the percentage of overlapping bars in the same section must be less than:

- 1/3 for tense sections with $M/N < 0.5\ h_o$,

- 1/2 in other cases.

The cover of the steel walls had to be at least:

– 5 cm in the case of salt fog, sea spray or sea water;

– 3 cm if the facings were exposed to the weather, in contact with a liquid, to condensation or depending on the aggressiveness of the environment;

– 1.5 cm inside buildings.

9.9.3. *Requirements for slabs and tank undersides resting directly on the ground*

In addition to the previous provisions, the thickness of the base must be greater than 0.10 m.

9.9.4. *Special provisions for domes*

The flexure of the covering and bottom domes of tanks was defined as a function of the diameter D of the tank by:

– $f \geq D/10$ for the covering dome;

– $f \geq D/8$ for the bottom dome (f is the height of the vault and D is its opening).

The thicknesses of the dome wall were given by:

– $h_o \geq 8$ cm for the covering dome;

– $h_o \geq 15$ cm for the bottom dome.

9.9.5. *Loads*

Loads	Nature	Value
Dead loads	– Dead weight of civil engineering – Dead weight of equipment – Weight of superstructures – Prestressing – Shrinkage – Earth weight and thrust – External weight and pressure Imposed displacements (e.g. differential settlements)	According to NFP 06.001 According to technical specifications According to NFP 06.001 According to BPEL
Variable loads	– Weight and pressure of the contained liquid – Operating loads – Climatic loads – Maintenance loads – Variations in water table height – Loads on the ground surface – Thermal gradient – Earth weight and pressure – Water table underpressure Water pressure on buried walls	Settling works: $\rho = 1{,}010$ daN/m^3 Lifting, pretreatment: $\rho = 1{,}050$ daN/m^3 Digesters: $\rho = 1{,}100$ daN/m^3 According to NFP 06.001 According to NV 65 modified in 1999 and N 84 modified in 2000 2 kN/m^2 for floors, walkways 1 kN/m^2 for terraces – Difference between PHE and PBE To be specified in the CCTP T_{max} and T_{min} per the CCTP (±20°C)
Accidental loads	– Earthquake Shocks	– According to PS 69 and PS 92 According to CCTP

Table 9.11. *Calculation of load values*

The loads to be considered when determining the structure sizes were those indicated in Table 9.11.

9.9.6. *Specific provisions*

– Covers: the values were identical to those of the 1983 version.

– Minimum reinforcement sections (non-fragility condition): the minimum sections A_s were fixed at:

- $A_s > 0.23$ ft28/fe in simple bending;
- $A_s > 0.23$ ft28/fe – $N/3h_o fe$ in compound bending;
- $A_s > f_{t28}/fe$ in simple traction.

Also:

– $A_s > 0.125\%$ concrete cross-section (for high-grip steel);

– $A_s > 0.20\%$ concrete cross-section (for smooth steel);

– $A_s < 2\%$ concrete cross-section.

The stresses for structural elements comprising tensioned sections or reinforcement close to the wetted walls of partially tensioned sections or reinforcement close to the wetted walls of fully tensioned sections were justified on the following basis:

– $\sigma_s < \alpha(\eta\ f_{t28}/\Phi)^{0.5} + \beta\eta$ (tensile strength of reinforcement);

– $\sigma_s < 90(\eta\ f_{t28})^{0.5}$;

– $\sigma_s < 0.5$ fe.

With:

– $\alpha = 240$ (except for laboratory-approved test results);

– Φ: diameter of the reinforcements (mm);

– f_{t28}: characteristic tensile strength of the concrete (MPa);

– $\beta = 0$ for structures at sea or near the sea (<5 km) as well as for buried parts of structures when the water is brackish or aggressive (wine tank, slurry pit, etc.);

– $\beta = 30$ for other cases of permanent contact with water or in a saturated atmosphere;

– η: reinforcement cracking coefficient;

– σ_s: tensile stress of steel.

For structural elements with fully tensioned sections or tensioned wet walls, the tensile stress of the concrete was limited to:

$$\sigma_b < 1.10\ \theta\ f_{t28} \text{ (concrete tensile strength)}$$

with:

– $\theta = 1$ (simple traction);

– $\theta = 1 + 2e_o/3h_o$ (compound bending);

– $\theta = 5/3$ (other cases).

9.10. Special case of silos: professional regulations of 1986

These regulations replaced those of 1975. This second edition differed from the previous one in the following aspects:

– The principles and methods of calculating the actions exerted by the ensiled material were confirmed by the observation of the behavior of many real cases.

– These rules (like the previous ones) only applied to powdery materials.

– The formulas applicable to silo-type structures have been extended to larger structures, and methods for calculating "storage warehouses" and "tank silos" have been introduced.

– The methods for determining stresses related to ovalization and the thermal gradient have been refined.

– The texts have been updated in relation to BAEL 83, which was applicable since January 1, 1985.

– Prestressed concrete silos were the subject of a specific chapter.

9.10.1. *Load combinations*

The following combinations were considered:

– for ULS:

$$(C_1) = 1.35(G) + 1.5(Q) + (W') + (T)$$

$$(C_2) = (G) + 1.5(Q) + (W') + (T)$$

$$(C_3) = (G) + 1.3(Q) + 1.5(W') + (T)$$

– for accidental ULS:

$$(C_4) = (G) + (Q) + (SI) + (T)$$

– for SLS (cracking):

$$(C_5) = (G) + (Q) + (W) + (T)$$

with:

– (G): dead loads;

– (Q): operating loads;

– (W) and (W′): climatic loads (in general, $W' = 1.2V$ and $W = V$, where V is the load of wind or snow S_n and $(1.12S_n - 60)$);

– (T): imposed deformations;

– (SI): earthquake loads.

In these combinations, the weight (Q) of the ensiled material was multiplied by the behavior coefficient k.

9.10.2. *Section supporting calculations*

The supporting calculations for the straight sections were carried out according to the BAEL 83 regulations, namely:

– for ULS, the stretched concrete was not considered, and the value of the stresses in the materials was limited to:

- $0.85f_{cj}/\gamma_b \times \psi_b$ for concrete (the value of ψ_b was 1 when the concrete temperature did not exceed 100°C; otherwise, it was appropriate to refer to the values defined in section 4.5.21 of the regulations),

- f_s/γ_s for steel (corresponding to 0.2% elongation);

– for SLS, the stresses were calculated with an equivalence coefficient n = 15. The compressive stress of the concrete was limited to $0.6f_{cj}$. The tensile stress in the vertical and horizontal reinforcements was limited to:

$$\bar{\sigma}_s = \alpha\,((\eta\ f_{t28})/\phi)^{1/2}\ (\text{MPa}) \leq 150\ \eta$$

with:

– η: reinforcement cracking coefficient;

– Φ: diameter of the frame in mm;

– f_{t28}: characteristic tensile strength of the concrete (in MPa);

– α: value depends on the wall category according to sections 4.4.23.2 and 4.4.23.3 (α = 400 for category 1 walls, 450 for category 2 walls, no value for category 3 walls).

9.10.3. *Storage warehouses and tank silos*

The regulations also dealt with storage warehouses and tank silos.

Given their low slenderness, the storage warehouses were considered to be retaining walls.

The calculation was carried out according to the emptying method, either from above (with a skip) or from below (through an underground gallery).

The values of the lateral pressures were given in the first case by:

– $n'_6 = \omega'_6 \gamma (z - h')$, where ω'_6 denotes the active thrust coefficient (straight line I) during the first emptying;

– $n''_6 = \omega''_6 \gamma z$ (straight line II) during the second emptying.

The calculation value was then deduced from this: $n = \text{Min}(n'_6; n''_6)$.

In the case of an underground gallery, the value of the lateral pressure was given by the previous expressions by replacing ω'_6 and ω''_6 with $\omega'_7 = \omega'_6 + \varphi/500$ and $\omega''_7 = \omega''_6 + \varphi/500$.

As for the tank silos, they were characterized by the following dimensions:

– $1.5 \leq H/r_H \leq 3.5$;

– $r_H \leq 7.5$ m;

– $h - h' \geq 0.6H$.

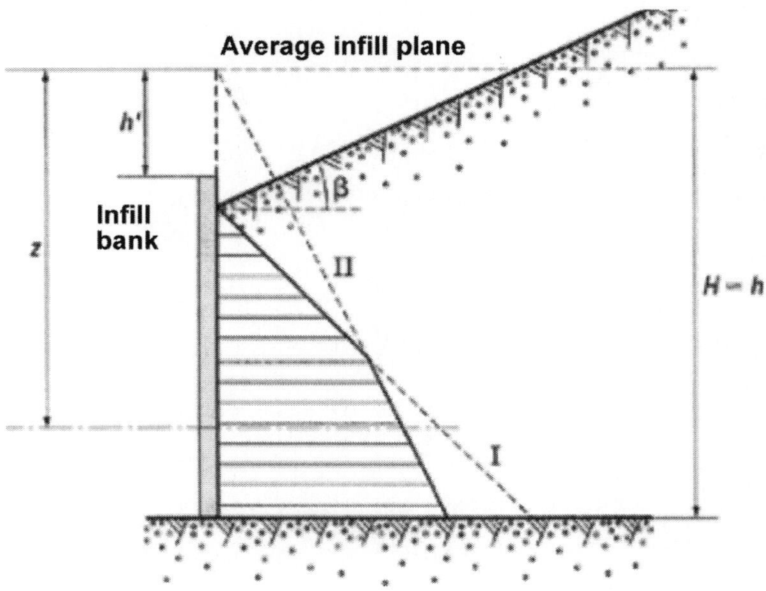

Figure 9.8. *Features of a storage warehouse*

The calculation was carried out as for a "normal" silo, but when the horizontal thrust n_c was evaluated, the coefficient 1.15 was replaced by:

– $k_{n1} = 1 + 0.075(H/r_H - 1.5)$ for state 1;

– $k_{n2} = k_{n1}\ (0.85 + 0.075\ (H/r_H - 1.5))$ for state 2.

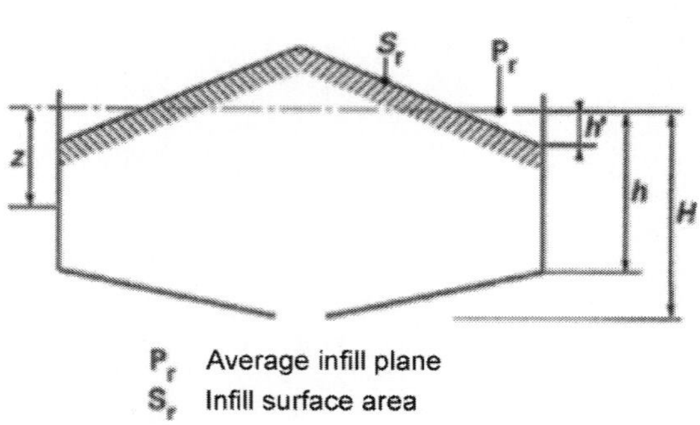

Figure 9.9. *Characteristics of a tank silo*

9.11. The 1969 earthquake regulations revised in 1982

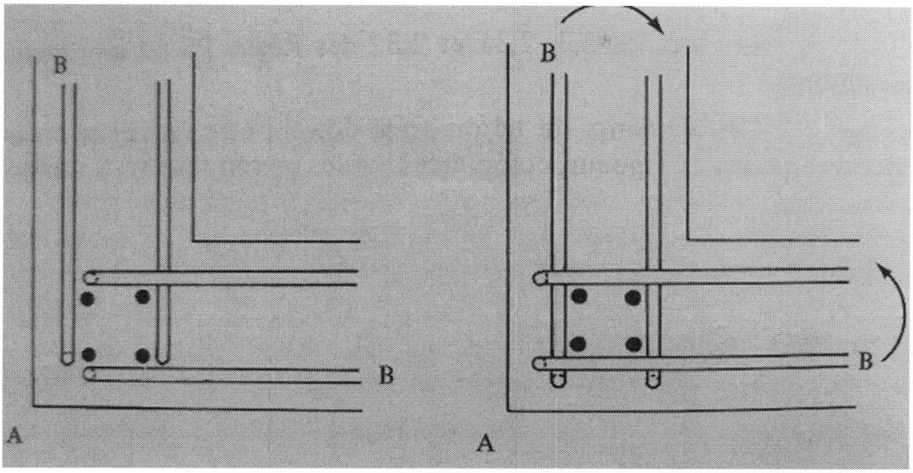

Figure 9.10. *Incorrect provisions*

The changes made in 1982 to the 1969 regulations did not fundamentally change the latter but were a complement pending the complete overhaul, which was carried out in 1992 with the publication of the PS 92 regulations.

The main objective of these additions was to reduce the risk of brittle or premature fractures in reinforced concrete elements.

They applied not only to ordinary reinforced concrete but also to reinforced lightweight aggregate concrete and concerned articles 2.3, 2.31, and 2.321 of the PS 69 regulations:

– Article 2.3: the elements of a frame had to constitute a mechanically continuous network in all directions.

– Article 2.31: in beams, an end region was considered in the vicinity of each node, defined as the part of the beam extending over a length equal to twice the total height of the latter, counting from the support. In columns, a critical length was considered in the vicinity of each knot. This length was defined as:

- the largest cross-sectional dimension;

- 1/6 of the free height;

- 450 mm;

- l/(2 (p – 3)) if the ratio p between the free height and the smallest dimension of the cross-section was less than 6 (l being the total height of the column).

The use of the hook was prohibited in knots, end regions and critical areas of the posts.

In these areas, the anchors had to be ensured by straight seals or regulatory right-angle returns.

In the end regions of the beams, the transverse reinforcements had to be such that the longitudinal steel parts were individually secured by a reinforcement opposing their buckling.

For columns, the smallest dimension had to be greater than 180 mm, and the ratio between the largest and smallest dimensions had to be less than 4.

The constructive provisions were modified as follows:

– Longitudinal reinforcement: the percentage of longitudinal reinforcement beyond overlapping areas had to be between 1% and 3%. The overlapping lengths had to be increased by 20%. There should not be more than half of the overlapping bars in the same area. The spacing of the longitudinal bars on the same face should not be greater than 200 mm.

– Transverse reinforcements: the diameter had to be greater than 6 mm. In critical areas, the percentage of transverse steels had to be greater than 1%, and the spacing was less than the Min ($8\phi_l$; a/2; 200 mm). With ϕ_l the diameter of the longitudinal reinforcements and a short side of the cross-section.

– Minimum dimensions: columns with sides less than 180 mm could not be considered in terms of resistance to lateral effects. However, their load-bearing function under downward vertical loads remains to be justified. With respect to the walls, the minimum thickness was 130 mm, and their width had to be greater than eight times the thickness and half the free height between floors; their cross-section was greater than 0.20 m^2.

– Expansion joint: the joints had to be purged with any material whose width was greater than 4 cm.

9.12. The 1991 amendments to the BAEL 1983 regulations

Decree No. 92–72 of January 16, 1992, made the application of the new CCTG booklet 62 (BAEL 91) mandatory. The main modifications were as follows:

– Application scope: extended to concrete with strengths greater than 40 MPa and up to 60 MPa.

– The formula for the mechanical resistance of concrete for values of j less than 28 days has been modified according to the requirements of article A.2.1, that is, $f_{cj} = j/(4.76 + 0.83j) \times f_{c28}$ for $f_{c28} \leq 40$ MPa and $f_{cj} = d/(1.40 + 0.95j) \times f_{c28}$ for $f_{c28} > 40$ MPa. The range of shrinkage values has been extended to all climates according to Table 9.12.

– Poisson's ratio was equal to 0 for the calculation of stresses and 0.2 for the deformation calculations.

– Width of compression tables for T-beams: Article A.4.1.3 removes the last paragraph and the corresponding comment.

– Stress–deformation diagram of concrete: the coefficient 0.85 was replaced by 0.85/θ to take into account the duration of load application. The values of θ were 1 when the application duration of the load combination was greater than 24 h, 0.9 when it was between 1 h and 24 h, and 0.85 for durations less than 1 h.

– Crack opening limit states: according to article A.4.5.3, for harmful and very harmful cracks, the values of η were replaced by $K (\eta f_{tj})^{1/2}$ with K = 110 in harmful cracking and K = 90 in very harmful cracking.

Location	Very humid climate	Humid climate	Dry temperate climate	Hot and dry climate	Very dry and desert climate
Value of the shrinkage coefficient	1.5×10^{-4}	2×10^{-4}	3×10^{-4}	4×10^{-4}	5×10^{-4}

Table 9.12. *Values of the concrete shrinkage coefficients*

The main differences from the 1983 version are summarized in Table 9.13.

NOTE.– These regulations were revised in 1999 to include high-performance concrete (HPC).

Articles of the Regulation	Article Contents	BAEL 83 version	BAEL 91 version
A.2.1.11	Value of f_{cj} for verifications	fc_{28} limited to 40 MPa	1.10 fc_{28}
A.2.1.11	Value of f_{cj} for $j \leq 28$ days	$0.685\ fc_{28} \log(j + 1)$	$j*fc_{28}/K$ with: $K = 4.76 + 0.83j$ for $fc_{28} \leq 40$ MPa $K = 1.40 + 0.95j$ for $fc_{28} > 40$ MPa
A.2.1.12	Tensile strength	$f_{tj} = 0.6 + 0.006f_{cj}$	$f_{tj} = 0.6 + 0.006f_{cj}$ whatever the value of f_{cj}
A.4.1.3	Compression table width	1/40 of the sum of the spans + 2/3 from the distance of the section to the axis of the support	Conditions removed
A.4.3.4	Design resistance of concrete σbc as a function of the duration of load application	$0.85\ f_{cj}/\gamma b$	$0.85\ f_{cj}/(\theta\ \gamma b)$ with: $\theta = 1$ if the duration > 24 h $\theta = 0.9$ if the duration is between 1 h and 24 h $\theta = 0.85$ if the duration $\leq$ 1 h
A.4.3.5	Eccentricity in compound bending	$e = e_1 + e_a + e_2$	$e = e'_1 + e_2$ where $e'_1 = e_1 + e_a$

Articles of the Regulation	Article Contents	BAEL 83 version	BAEL 91 version
A.4.4.3	Calculation inclination	0.01 rd	0.01 rd for one floor 0.005 rd for other frames
A.4.5.33/34	Cracking at SLS: Damaging cracking: σ_s = Very damaging cracking: σ_s = (η = 1.6 for high-grip steel and TS with $\phi \leq 6$ mm, η = 1.3 for TS with ϕ > 6 mm)	Min (2/3 fe; 150η) MPa Min (0.5fe; 110η) MPa	Min (2/3 fe; 110 (η $f_{tj})^{1/2}$) MPa Min (0.5 fe; 90 (η $f_{tj})^{1/2}$) MPa
A.5.1.1	Shearing of compressed parts. The requirements of the following § were not to be applied if: $\tau_u = V_u/(b_0d) \leq$	Min (0.04 f_{cj}; 1.5 MPa)	Min (0.06f_{cj}/γ_b; 1.5 MPa)
A.5.1.21	ULS of the core concrete. Limitation of shear stress $\tau_u = V_u/(b_0d)$ When: The reinforcement bars are straight and the cracking is minimally harmful: The cracking is harmful or very harmful: The reinforcement bars are inclined at 45°:	 Min (0.13 f_{cj}; 4 MPa) Min (0.10 f_{cj}; 3 MPa) Min (0.18 f_{cj}; 5.5 MPa)	 Min (0.20 f_{cj} / γ_b; 5 MPa) Min (0.15 f_{cj} /γ_b; 4 MPa) Min (0.27 f_{cj} /γ_b; 7 MPa)

Articles of the Regulation	Article Contents	BAEL 83 version	BAEL 91 version
A.5.1.23	Limit state of the web reinforcement: In the vicinity of the support for the calculation of the web reinforcement, the distributed loads can be neglected up to h/2: Taken in fraction 2a/3 h up to 1.5 h: The cross-section of the core reinforcements A_t is such that $A_t/(b_0$ st) =	Yes No $(\tau_u - 0.3k\ f_{tj})/(0.8\ fe\ (\cos\alpha + \sin\alpha))$ k = 0 in the event of concrete rework or very damaging cracking	Yes Yes $\gamma_s(\tau_u - 0.3kf_{tj})/(0.9\ fe\ (\cos\alpha + \sin\alpha))$ f_{tj} being limited to 3.3 MPa k = 0 in the event of concrete rework or very damaging cracking k = 1 if the recovery surface is indented at least 5 mm
A.5.2.2	Justification of shear reinforcement No reinforcement is required if:	The room is concreted without rework and $\tau_u \leq 0.05\ f_{cj}$	The room is concreted without rework and $\tau_u \leq 0.07\ f_{cj}/\gamma_b$
A.5.2.4	Punching under Q_u No shear force reinforcement is required if:	$Q_u \leq 0.045u_c\ h\ f_{cj}$	$Q_u \leq 0.045u_c\ h\ f_{cj}/\gamma_b$
A.5.4.3	Torsion Torsional shear stress τ_T and shear force τ_v are such that: Hollow sections: Solid sections:	$\tau_T + \tau_v \leq \tau_{lim}$ $\tau_T + \tau_v \leq \tau_{lim}$	$\tau_T + \tau_v \leq \tau_{lim}$ $\tau_T^2 + \tau_v^2 \leq \tau_{lim}^2$

<table>
<tr><th>Articles of the Regulation</th><th>Article Contents</th><th>BAEL 83 version</th><th>BAEL 91 version</th></tr>
<tr><td rowspan="4">A.7.1</td><td>Coating:
Structures at sea or exposed to sea spray:</td><td>4 cm</td><td>5 cm or 3 cm if protection</td></tr>
<tr><td>Unformed facings subject to aggressive loads:</td><td>3 cm</td><td>3 cm or 2 cm if $f_{c28} \geq 40$ MPa</td></tr>
<tr><td>Facings exposed to bad weather or water loads:</td><td>2 cm</td><td>3 cm or 2 cm if $f_{c28} \geq 40$ MPa</td></tr>
<tr><td>Walls of covered and enclosed spaces without condensation:</td><td>1 cm</td><td>1 cm</td></tr>
<tr><td>A.8.2.42</td><td>Spacing of steels in the same layer in the direction most stressed under concentrated load:</td>
<td><table>
<tr><th>Directions</th><th>Distributed charges only</th><th>Concentrated charges</th></tr>
<tr><td>Most stressed direction</td><td>3 h and 33 cm</td><td>2 h and 22 cm</td></tr>
<tr><td>Direction perpendicular to the most stressed</td><td>4 h and 45 cm</td><td>3 h and 33 cm</td></tr>
</table></td>
<td><table>
<tr><th>Directions</th><th>Distributed charges only</th><th>Concentrated charges</th></tr>
<tr><td>Most stressed direction</td><td>3 h and 33 cm</td><td>2 h and 25 cm</td></tr>
<tr><td>Direction perpendicular to the most stressed</td><td>4 h and 45 cm</td><td>3 h and 33 cm</td></tr>
</table></td></tr>
<tr><td>A.8.3.1</td><td>Localized pressures at ULS:
Case of concentric rectangles:

Case of homothetic surfaces:</td>
<td>$\sigma_{bc} = K \times 0.85 \times fcj/\gamma_b$
$K = 1 + (3 - a_0/a - b_0/b)\ ((1 - a_0/a)\ (- b_0 / b))^{1/2}$

$K = 4 - 5(B_0/B)^{1/2} + 2B_0/B$</td>
<td>$\sigma_{bc} = K \times 0.85 \times fcj / \gamma_b$
$K = 1 + (3 - 4 / 3(a_0/a + b_0/b))\ ((1 - 4a_0/3a)\ (1 - 4b_0/3b))^{1/2} \leq 3.3$
$K = 4 - 20/3(B_0/B)^{1/2} + 32B_0/9B \leq 3.3$</td></tr>
<tr><td>B.2.4</td><td>Exposure characteristics</td><td>§ A.4.5.33</td><td>§ B.2.4</td></tr>
</table>

Articles of the Regulation	Article Contents	BAEL 83 version	BAEL 91 version
B.6.5.2	Evaluation of flexure If: cracked inertia I0: homogeneous section inertia λ and μ coefficients defined in the text	Calculation in cracked inertia with: $I_f = I_0/(1 + \lambda\ \mu)$	Calculation in cracked inertia with: $I_f = 1.1\ I_0/(1 + \lambda\ \mu)$
B.6.6.2	Skin reinforcement In minimally damaging cracking: In damaging cracking: In very damaging cracking:	§ A.5.3 3 cm^2/m 3 cm^2/m 5 cm^2/m	(beams such that $h \leq 25$ cm and $b_0 > 15$ cm) 0 1 cm^2/m 5 cm^2/m
B.6.7.1	Transverse reinforcement. Exception for secondary beams and ribs of cross-ribbed floors. No transverse reinforcement if: Unreinforced length: For secondary beams: For floor ribs:	$\tau_u \leq 0.025\ f_{c28}$ (if no dynamic effects and $h \leq 0.40$ m) 0.5 L central 0.5 L central	$\tau_u \leq 0.0.03\ f_{c28}$ (if no dynamic effects) Max (0.5 L; L – 4 h) L

Articles of the Regulation	Article contents	BAEL 83 version	BAEL 91 version
B.6.7.2	Table-rib connection. For moderately loaded floors, The stress of the seam steels is not verified if:	$\tau_{umax} \leq 0.025\ f_{c28}$	$\tau_{umax} \leq 0.05\ f_{c28}$ if no vertical recovery $\tau_{umax} \leq 0.025\ f_{c28}$ with vertical recovery
	The sliding force is considered uniform if:	$\tau_{umax} \leq 0.05\ f_{c28}$	$\tau_{umax} \leq 0.10\ f_{c28}$ if no vertical recovery $\tau_{umax} \leq 0.05\ f_{c28}$ with vertical recovery
B.6.8.3	Beams supporting walls The maximum compressive stress at ULS in relief vaults is: In minimally damaging and damaging cracking: In highly damaging cracking:	2 MPa for masonry $f_{c28}/3$ for concrete 2 MPa for masonry $f_{c28}/3$ for concrete	2 MPa for masonry $f_{c28}/2$ for concrete 2 MPa for masonry $f_{c28}/3$ for concrete

Table 9.13. *Comparison of BAEL 83 and 91 regulations*

9.13. The 1995 amendments to the PS regulations

In December 1995, the PS 92 regulations (NFP 06.013) were published. Compared with the PS 69 regulations revised in 1982, the differences concerned the following points:

– consideration of geotechnical problems such as soil liquefaction, slope instabilities, local topographic amplification, soil/structure interaction and the particularities of buried levels;

– more demanding construction provisions;

– an incursion into the postelastic domain through a behavior coefficient q;

– the adaptation of calculations according to structural regularity;

– the increase in stresses for formwork concrete walls and masonry walls;

– the penalizing effect for structures with low ductility and bracing by rigid walls.

These regulations applied to:

– normal risk buildings only (excluding special risk buildings);

– traditional construction materials and systems (excluding those not covered by normative texts).

The definition of earthquake loads was given by the use of the $R_D(t)$ spectrum when sizing buildings (section 5.2.3). This was expanded to consider the topographic amplification coefficient ST and the corrective damping coefficient ρ. The determination of the sizing spectrum was made from the parameters R_M, T_C and T_D.

The classification of soils into four groups was carried out based on mechanical properties (healthy rock and groups a–c) and allowed classification into four types of sites (S_0, S_1, S_2 and S_3) depending on the soil group and soil thickness.

The nominal design acceleration was then a function of the earthquake zone and the class of the structure according to Table 9.15.

The value of the nominal acceleration a_N is then given in Table 9.16.

Soil type		Static penetrometer Resistance (MPa)	SPT number of blows	Pressure meter		Resistance	Relative density	Compression index	Shear wave velocity	Longitudinal wave velocity	
				Modulus (MPa)	Ultimate pressure (MPa)	Compression simple (MPa)	(%)	(C_c)	(m/s)	Below the water table (m/s)	Above the water table (m/s)
Rocks	Healthy rocks and hard chalk			> 100	> 5	> 10			> 800		> 2 500
Category A Soils with good to very good volcanic resistance	Compacted granular soils	> 15	> 30	> 20	> 2		> 60		> 400	> 1 800	> 800
	Coherent soils (hard clays or marls)	> 5		> 25	> 2	> 0,4		< 0,02			> 1 800
Category B Soils with medium volcanic resistance	Weathered rocks or fractures			30 to 100	2.5 to 5	1 to 30			300 to 800		400 to 2 500
	Medium-compacted granular soils	5 to 15	10 to 30	6 to 20	1 to 2		40 to 60		150 to 400	1 500 to 1 800	500 to 800
	Medium-consistent coherent soils and soft chalks	1.5 to 5		5 to 75	0.5 to 2	0.1 to 0.4		0.02 to 0.10			1 000 to 1 800
Category C Soils with low volcanic resistance	Loose granular soils	< 5	< 10	< 6	< 1		< 40		< 150	< 1 500	< 500
	Soft coherent soils (soft clays or silts) and weathered chalks	< 1.5	< 2	< 5	< 0.5	< 0.1		> 0.10			

Table 9.14. *Table of soil categories*

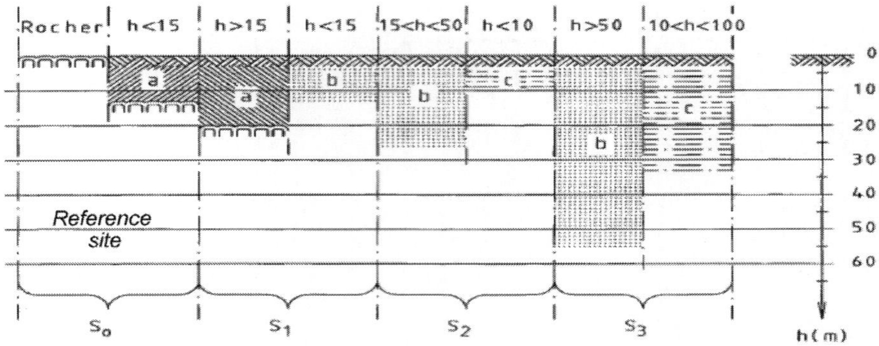

Figure 9.11. *Classification of sites*

Class	Building characteristics	Examples
A	A structure whose failure represents only minimal risk to people or the economy.	Buildings in which any human activity requiring a long stay is excluded.
B	Structures and installations presenting a so-called normal risk for the population.	Individual residential buildings. Multi-family residential or office buildings with a height not exceeding 28 meters. 4th and 5th category public buildings. Buildings housing public parking lots. Other buildings that can accommodate fewer than 300 people (non-public buildings, industrial activities).
C	Structures representing an increased risk due to their use or their socioeconomic importance	Multi-family residential or office buildings over 28m. Public buildings of the 1st, 2nd, and 3rd categories. Other buildings that can accommodate more than 300 people. Health and social buildings, with the exception of those listed in Class D.
D	Structures and buildings in which safety is important	Buildings whose protection is essential for the needs of civil security and national defense and the maintenance of public order (operational resources, defense buildings) Buildings that contribute to maintaining communications (telecommunications centers, broadcasting and reception centers, relays, airport control towers, air traffic control rooms) Public hospital buildings that provide short-term care, surgery, and obstetrics *Drinking water production and storage buildings* Public energy distribution centers Meteorological centers

Table 9.15. *Table of building classes*

Class of structures/earthquake zones	A	B	C	D
0				
Ia		1.0	1.5	2.0
Ib		1.5	2.0	2.5
II		2.5	3.0	3.5
III		3.5	4.0	4.5

Table 9.16. *Nominal acceleration values (in m/s²)*

NOTE.– In Zones Ia and Ib, earthquake joints between blocks of at least 4 cm were recommended; in Zones II and III, this value increased to 6 cm.

We then deduced that the value of the earthquake load R(t) was given by:

$$R(t) = a_N \, R_D(T) \, \rho \, ST$$

The normalized spectra were given for a value of 5% relative damping. Their general form is given in Figure 9.12.

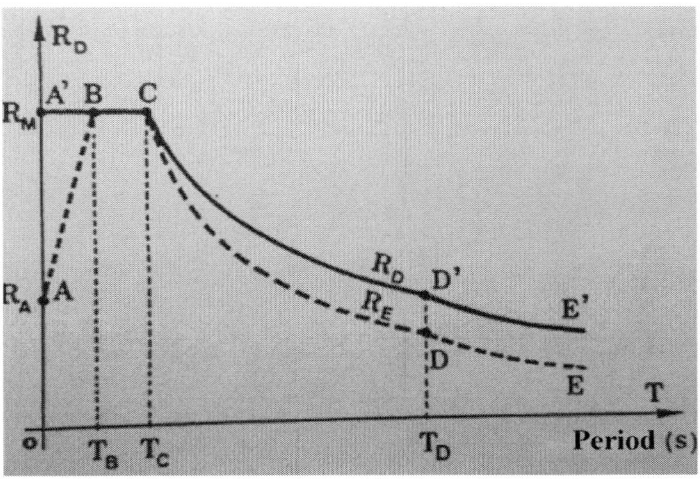

Figure 9.12. *Normalized sizing spectrum*

The spectrum equation responded to the analytical form in which the ordinates are dimensionless numbers:

– branch A′C: $R_D(T) = R_M$;

– branch CD′: $R_D(T) = R_M\ (T_C/T)^{2/3}$;

– branch D′E: $R_D(T) = R_M\ (T_C/T_D)^{2/3}\ (T_C/T)^{2/3}$.

The sizing spectra were derived from the normalized elastic spectra by replacing the ascending branch AB of the latter by a horizontal step extending step BC and by raising the ordinates of the descending branches.

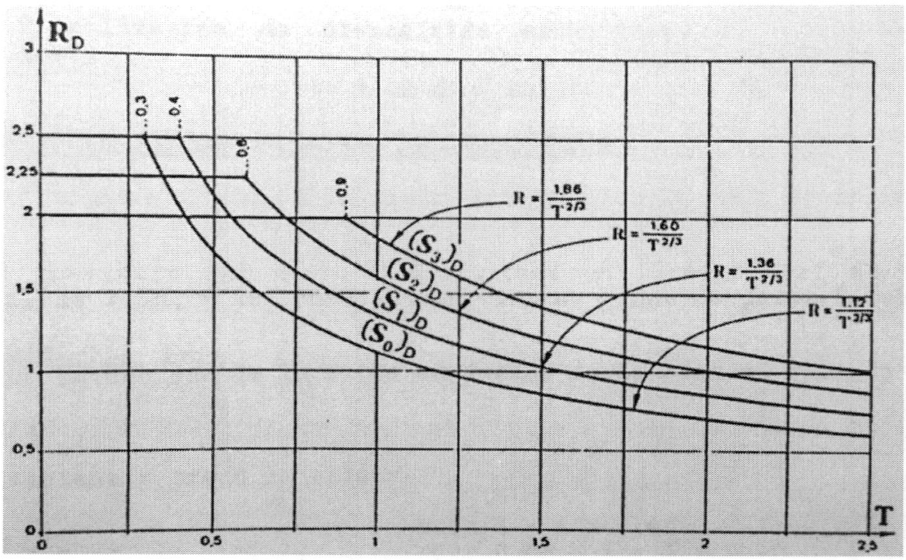

Figure 9.13. *Horizontal spectrum component*

The spectrum of the vertical component was considered to be identical to the spectrum (S_0) or (S_1); in other cases, the descending branches of the spectrum were replaced by those of the spectrum (S_1). Then, an affinity ratio of 0.7 was introduced.

When the damping differed from 5%, a coefficient ρ was adopted such that:

$\rho = (5/\xi)^{0.4}$, where ξ is the damping calculation

The topographic amplification coefficient was determined as a function of the slope of downstream slope I and that of the upstream slope for ridge heights above the relief base greater than 10 m, such that:

– ST = 1 for $I - i \leq 0.40$;

– ST = $1 + 0.8 (I - i - 0.40)$ for $0.40 \leq I - i \leq 0.90$;

– ST = 1.40 for $I - i > 0.90$.

Chapter 11 of the regulations dealt with the specificities of reinforced concrete and prestressed concrete structures and specified:

– what the primary and secondary elements of structures were;

– the location of critical zones (parts of a main structural element in which concentrations of strains or stresses are likely to occur);

– the concept of confinement: the volume of concrete provided with transverse reinforcements arranged in such a way as to oppose the swelling of the latter under the effect of compressive stresses as well as the buckling of the longitudinal reinforcements;

– the concept of reduced normal effort: $\upsilon = N_d/(B_n \times f_{cj})$, where N_d is the normal calculation effort, B_n is the net cross-section of the concrete and f_{cj} is the compressive strength of the concrete;

– the concept of compressed and bent parts: a part became bent when $\upsilon_{max} \leq 0.07$ and compressed when $\upsilon_{max} > 0.17$.

With respect to the materials, the regulations applied to concrete with a mechanical strength between 25 and 45 MPa. For the main elements, the reinforcements were high-grip steel with an elastic limit less than or equal to 500 MPa with a relative elongation greater than or equal to 5%.

With respect to the construction arrangements, the end anchors were built as shown in Figure 9.14.

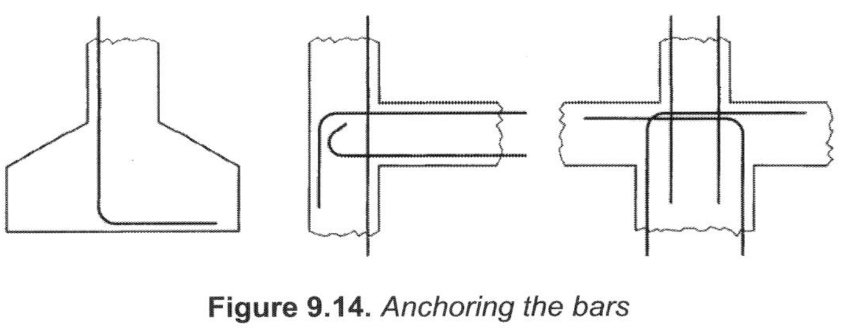

Figure 9.14. *Anchoring the bars*

The transverse reinforcements of the critical zones were shaped according to the arrangements shown in Figure 9.15.

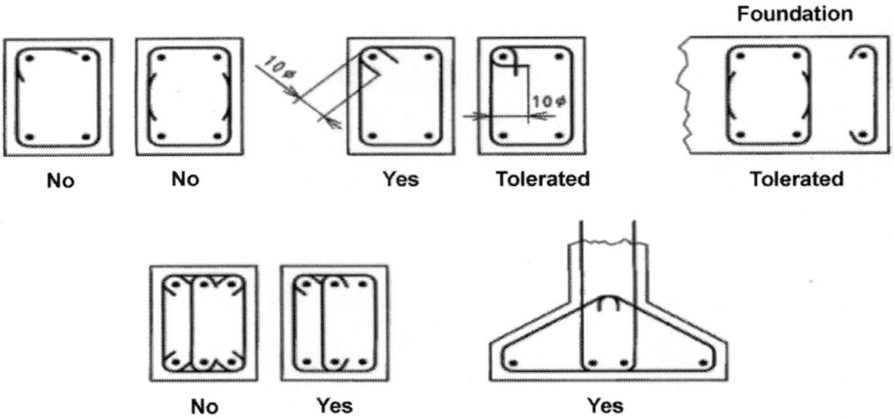

Figure 9.15. *Constructive provisions of transverse reinforcements*

9.13.1. *Provisions common to beams and posts*

The linear elements had to have minimum cross-sections a and b such that:

– a > 0.25 m and b > 0.25 m;

– cross-section B = a × b > 6.25 cm²;

– the beam core thickness was greater than 0.15 m.

9.13.2. *Specific provisions for bent elements*

The critical zones of the bent elements were defined by a length l_{crit} (see Figure 9.16).

The geometric percentage ρ_0 of longitudinal reinforcements arranged on the tensioned side had to satisfy the following conditions:

– $\rho_{0min} > 1.4/fe$;

– $\rho_{0max} \leq 0.025$.

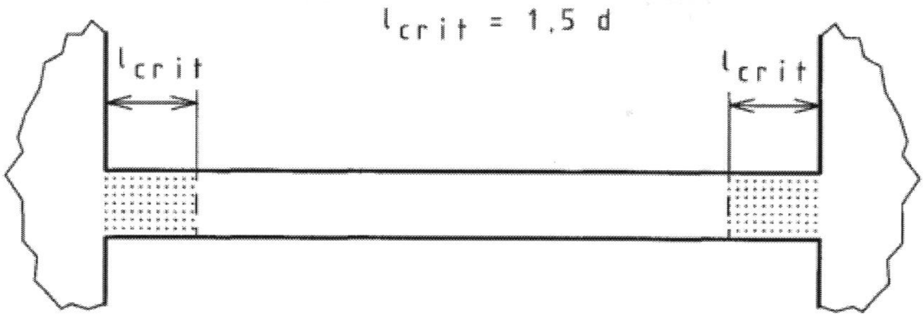

Figure 9.16. *Definition of length from the useful height d*

In addition, it was required that a quarter of the reinforcements placed at the ends be continuous over the entire length of the part and that in the critical zones, the cross-section of the compressed reinforcements be at least equal to half the cross-section of the tensioned reinforcements.

In critical areas, the shear force reinforcements had to be such that:

– their diameter is greater than 6 mm;

– their spacing is less than Min ($24\phi_t$; $8\phi l$; 0.25d), where ϕ_t is the diameter of the transverse reinforcements, ϕ_l is the diameter of the longitudinal reinforcements and the useful height.

9.13.3. *Specific provisions for compressed elements*

The critical areas were defined as in Figure 9.17.

The percentage of longitudinal reinforcements ρ_0 (beyond the overlap areas) had to be between 1% and 5%, and their spacing was less than 25 cm.

The transverse reinforcements had to satisfy the following conditions:

– minimum diameter: 8 mm;

– in critical areas, a minimum volume of 0.8% with spacing less than Min (0.25a; $8\phi_l$; 15 cm);

– in common areas, spacing was limited to the Min (0.5a; $12\phi_l$; 30 cm).

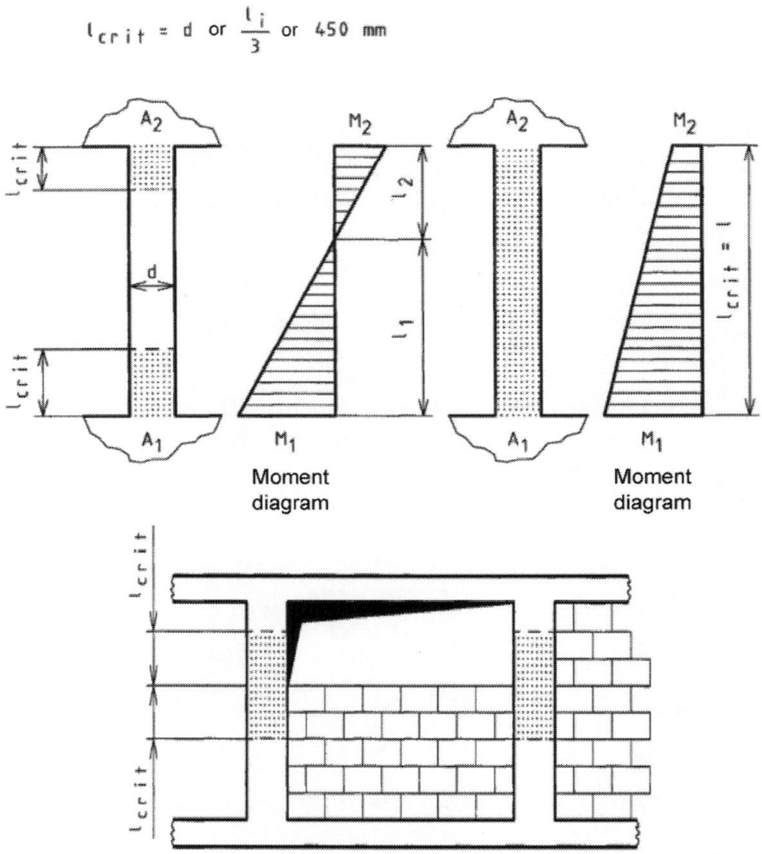

Figure 9.17. *Definition of critical zones of compressed elements*

9.13.4. *Knot-specific provisions*

The volume of the transverse reinforcements had to correspond to that of the compressed elements leading to the knot.

9.13.5. *Specific provisions for walls*

The walls had to have a minimum thickness of 15 cm and a width greater than four times the thickness.

A vertical chain was placed at each end of the wall as well as at all openings and intersections. It had to be continuous over the entire height of the floor and covered from floor to floor with stitching steel at the overlaps. In the current zone, the chains had to be made up as follows:

– vertical chaining: 4 HA10 with HA6 spaced 10 cm apart maximum;

– horizontal chaining: 2 HA10.

In the critical zone, the provisions concerning vertical chains were increased to 4 HA12 with transverse reinforcements in HA6 every 10 cm. These provisions were also valid for changes to the wall cross-section.

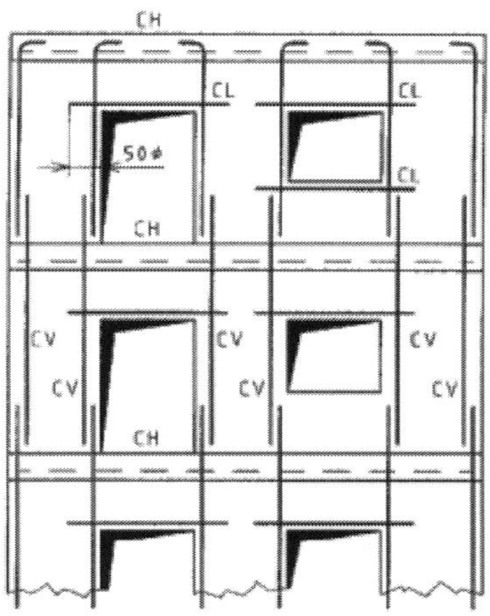

Figure 9.18. *Example of a chain arrangement*

9.13.6. *Specific arrangements for the slabs and diaphragms*

Each slab had to have a continuous peripheral chain of at least 3 cm² and a chain at each intersection with a shear wall of 1.5 cm².

9.13.7. *Behavioral coefficient*

In the absence of justifications, the values of the behavior coefficients q are mentioned in Table 11 of the regulations (Table 9.17).

	Types of structures	Regular buildings q_i	Moderately regular buildings	Irregular buildings
1	Structures braced only by walls	3.5 *)	0.85 q_1 *)	0.7 q_1 *)
2	Structures braced only by frames	5	0.85 q_2	0.7 q_2
3a	Chained load-bearing masonry	2.5	0.85 q_3	0.7 q_3
3b	Reinforced and chained load-bearing masonry	3	0.85 q_3	0.7 q_3
3c	Framework with post-infill	1.5	0.85 q_3	0.7 q_3
4	Mixed structures braced only by walls, frames, or masonry	Interpolation between the values 1, 2, and 3 by the formula $\frac{1}{q} = \sqrt{\frac{\sum\left(\frac{V_i}{q_i}\right)^2}{\sum V_i^2}}$ (curve given in Figure 60)		
5	Structures operating as a vertical cantilever with predominantly distributed masses	3	0.05 q_5	0.7 q_5
6	Structures with transparencies	Not applicable	2 to 3*)	1.5 to 2.5*)
*)	*Structures with transparencies and structures braced by sails must be subject to deformation compatibility verification (see 11.8.2.3)*			

Table 9.17. *Values of the behavior coefficient q*

9.13.8. *Safety check*

The characteristic diagrams were those of BAEL. The partial safety factors were 1 for steel and 1.15 for concrete. Verifications related to the shear limit force were carried out in accordance with BAEL with partial safety factors of 1 for steel and 1.15 for concrete but with an additional safety factor of 1.25.

In common areas and in the absence of reconcreting, the shear stress resisting the ultimate limit state τ_u must verify:

$$\tau_u \ll (0.8 f_e \left(\frac{A_t}{b_0 S_t}\right) + 0.3 f_{tj} 1/1.25$$

In critical zones, the shear stress at the ultimate limit state t_u must verify:

$$t_u \ll \left(0.8 f_e \left(\frac{A_t}{b_0 S_t}\right)\right) 1/1.25$$

For wall verification, reference was made to DTU 23.1.

10

European, American and Russian Normative Codes

10.1. European codes (Eurocodes)

10.1.1. *Overview*

The Eurocodes are the most recent codes for the design and calculation of structures. Eurocode 2 (reinforced concrete and prestressed concrete) replaces the BAEL and BPEL regulations. Prior to the final publication of Eurocode 2 as a standard (EN 1992), a provisional version was published in the form of an experimental standard (ENV 1992). The latter was published by AFNOR in 1992 with its national application document (NAD); it was then inspired by the European model code *CEB-FIP model code 1990*. During this transitional period, France invalidated a certain number of articles, particularly on shearing and constructive provisions, and renewed the articles of the BAEL in the form of NADs.

The transition from ENV to EN was completed in 2004. AFNOR then published the final version of EC 2 in the form of an NF standard and under the name NF-EN 1992.1.1 in October 2005. Moreover, the French Commission published recommendations in 2013 under the title "Guide to the application of Eurocode 2". In 2027, a complete revision of the European normative codes will be carried out.

The purpose of this chapter is not to be a course on applying Eurocode 2; there is already much literature on this subject, among which we can indicate:

– the calculation of concrete structures by Jean-Marie Paillé;

– Jean Roux's Practice of Eurocode 2;

– Aide-mémoire: reinforced concrete works by Arnaud Bury.

10.1.2. *Eurocodes calculation bases*

The regulation defining the bases for calculating structures, NF EN 1990, was published for the first time in March 2003. This code introduces a new concept, that of the duration of use of structures (Table 10.1).

The calculation principles are identical to those previously described for the BAEL regulations; thus, we find the concepts of ultimate limit states (ULSs) for structural safety and the safety of people, the serviceability limit states (SLS) for structural functionality, and the comfort of use and appearance. This last constraint is new and did not appear in the BAEL regulations. Chapter 7 of EN 1992.1.1 thus specifies the information in Figure 10.1.

Like the BAEL regulations preceding it, EN 1990 distinguishes three classes of actions:

– permanent actions (G): own weight, fixed equipment, indirect actions (shrinkage, differential settlements);

– variable actions (Q): operating costs, climate actions;

– accidental actions (A): explosions, fire, shocks and avalanches.

NOTE.– Seismic actions can be considered accidental or variable actions.

Actions are defined in characteristic values.

The action values are given in EN 1991. They are defined as characteristic values of uniformly distributed loads q_k and concentrated loads Q_k depending on the structure's categories of use. EC 1 distinguishes four categories of use:

– Category A, which covers residential buildings;

– Category B, which covers offices;

– Category C, which covers meeting places;

– Category D, which covers businesses;

– Category E, which covers storage areas and industrial premises;

– Categories F and G, which cover parking lots;

– Categories H, I, and K, which cover roofs.

The National Annex defines the values of Tables 10.2 and 10.3 for operating loads.

Project service life category	Indicative project service life (years)	Examples
1	10	Temporary structures[a)]
2	10 to 25	Replaceable structural elements, e.g., rolling beams, bearings
3	15 to 30	Agricultural and similar structures
4	50	Building structures and other common structures
5	100	Monumental structures of buildings, bridges, and other civil engineering works

a) Structures or parts of structures that can be dismantled for reuse should not normally be considered temporary.

Table 10.1. *Project usage duration according to EN 1990*

7.3 Cracking control

7.3.1 General considerations

(1)P Cracking must be limited so that it does not impair the proper functioning or durability of the structure or render its appearance unacceptable.

Figure 10.1. *Excerpt from section 7.3.1 of EN 1992.1.1*

Loaded area category	q_k [kN/m²]	Q_k [kN]
Category A		
– Floors	1.5 to 2.0	2.0 to 3.0
– Stairs	2.0 to 4.0	2.0 to 4.0
– Balconies	2.5 to 4.0	2.0 to 3.0
Category B	2.0 to 3.0	1.5 to 4.5
Category C		
– C1	2.0 to 3.0	3.0 to 4.0
– C2	3.0 to 4.0	2.5 to 7.0 (4.0)
– C3	3.0 to 5.0	4.0 to 7.0
– C4	4.5 to 5.0	3.5 to 7.0
– C5	5.0 to 7.5	3.5 to 4.5
Category D		
– D1	4.0 to 5.0	3.5 to 7.0 (4.0)
– D2	4.0 to 5.0	3.5 to 7.0

Table 10.2. *Operating loads according to the European standard EN 1991*

NOTE.–

– According to the European standard, Q_k is something new. This load must be applied at any point on the floor, on an impact surface of 50 mm × 50 mm.

– It is possible to apply a reduction coefficient for the surface effect. Its value is defined in the national appendix and can only be used for usage categories A, B, C_3, D_1 and F (Figure 10.2).

The characteristic values of the actions are defined, as in the BAEL regulations, from the main values Q_k and combination values according to the following formulas:

– combined value: ψ_0 Q_k;

– frequent value: ψ_1 Q_k;

– quasipermanent value: ψ_2 Q_k.

Loaded area category	q_k [kN/m²]	Q_k [kN]
Category A		
— **Floors**	1.5	2.0
— **Stairs** (1)	2.5	2.0
—Balconies	3.5	2.0
Category B	2.5	4.0
Category C		
— C1	2.5	3.0
— C2	4.0	4.0
— C3	4.0	4.0
— C4	5.0	7.0
— C5	5.0	4.5
Category D		
— D1	5.0	5.0
— D2	5.0	7.0
(1) except for independent markets, which are part of a dynamic approach		

Table 10.3. *Operating loads according to the French national appendix*

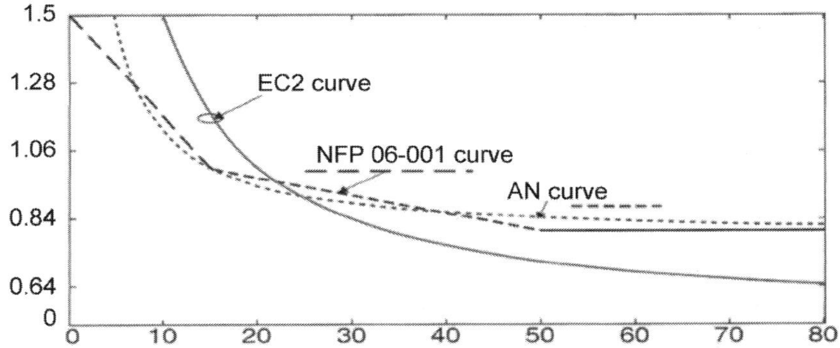

Figure 10.2. *Load reduction coefficients according to the different standards*

The coefficients ψ are specified in Table 10.4.

Action	ψ_0	ψ_1	ψ_2
Building operating load (see EC1, Part 1-1)			
Category A : Residential areas	0.7	0.5	0.3
Category B : Offices	0.7	0.5	0.3
Category C : Meeting areas	0.7	0.7	0.6
– Category D : Shops	0.7	0.7	0.6
– Category E : Storage	1.0	0.9	0.8
– Category F : Traffic area, vehicles weighing < 30 kN	0.7	0.7	0.6
– Category G : Traffic area, vehicles weighing between 30 and 160 kg	0.7	0.5	0.3
– Category H : Roofs	0	0	0
Snow loads on buildings in mainland France (see EC1, Part 1-3)			
– For locations located at an altitude H > 1,000 m and for Saint Pierre and Miquelon	0.7	0.5	0.2
– For locations located at an altitude H < 1000 m	0.5	0.2	0
Wind loads on buildings (see EC1, Part 1-4)	0.6	0.2	0
Temperature (non-fire) in buildings (see EC1, Part 1-5)	0.6	0.5	0

Table 10.4. *Coefficient ψ values*

The action combinations are identical to those of the BAEL regulations.

A new feature of the Eurocodes lies in the appearance of the Anglo-Saxon concept of *consequence classes* and *reliability classes*. These differentiations are expected to be made at the safety level of the structures according to the issues they represent.

The Eurocodes have thus introduced the principle of risk analysis as a method for evaluating the project owner choices. This method provides a hazard/consequence matrix analysis. It responds to the principles of precaution and sustainability management.

As a result, the project owner must first define the class of consequences of the project to determine its expected reliability. The latter calculates the probability of non-ruination under the actions taken into account during its service life. The classes of consequences are defined in EN 1990 and shown in Table 10.5.

The Eurocode then proposes several methods for defining reliability. There are three reliability classes (RC1, RC2 and RC3) that can be associated with the three consequence classes CC1, CC2 and CC3. The reliability index β is then defined in Table 10.6.

Consequence class	Description	Examples of buildings and civil engineering works
CC3	**High** consequence in terms of loss of life, or very significant economic, social, or environmental consequences	Grandstands, public buildings where the consequences of failure would be high (e.g., concert hall)
CC2	**Medium** consequence in terms of loss of life, or considerable economic, social, or environmental **consequences**	Residential and office buildings, public buildings where the consequences of failure would be medium (e.g., office building)
CC1	**Low** consequence in terms of loss of life, or low or negligible economic, social, **or environmental consequences**	Normally unoccupied agricultural buildings (e.g., storage buildings, greenhouses)

Table 10.5. *Definition of confidence classes*

Weakness class	Minimum values for β	
	1 year reference life	50 year reference life
RC3	5.2	4.3
RC2	4.7	3.8
RC1	4.2	3.3
Note: A design using EN 1990 with the partial factors given in Annex A1 and EN 1991 is generally considered to result in a structure rated at 3.8 for a reference life of 50 years. Weakness classes for structural elements higher than RC3 are not discussed further in this annex, as these structures require special consideration.		

Table 10.6. *Definition of reliability coefficient values*

Project supervision levels	Characteristics	Recommended minimum requirements for the control of calculations, drawings, and specifications
DSL321 associated with RC3	Enlarged supervision	Control by third party Control by an organization different from the one that prepared the project
DSL321 associated with RC2	Normal supervision	Control by people different from those initially responsible and in line with the organization's procedure
DSL321 associated with RC1	Normal supervision	Self-control Control by the person who prepared the project

Table 10.7. *Definition of the design supervision level (DSL)*

Depending on the reliability class chosen, EN 199 then imposes a design supervision level according to Table 10.7.

10.1.3. *Structural analysis*

EN 1992 introduced the concept of structural analysis. The aim of this analysis is to determine the distribution of stresses and deformations of a structure. The Eurocode is based on four behavior models:

– Elastic behavior: this is the principle of linear elastic analysis that can be used in the ULS and serviceability limit state (SLS). The model follows the RDM requirements. An initial stiffness corresponding to an uncracked inertia of the concrete is taken into account.

– Elastic behavior with limited redistribution: analysis limited to ULS. The model follows that of linear elasticity but with limited redistribution of moments on the supports. This redistribution assumes that the critical sections have sufficient rotation capacity to accept it.

– Plastic behavior: it uses kinematic and upper bound methods and is applicable only to the ULS. This method was not retained by the BAEL regulations. It assumes that the sections are sufficiently ductile to consider the creation of plastic hinges.

– Nonlinear behavior: it uses the curvature integration method and can be used for the SLS and ULS. The analysis is performed on the basis of a characteristic curve deduced from the RDM by translation, which allows a moment M to be calculated for each section of the structure. A classic bending calculation is then carried out for the ULS.

In addition, the ULS calculation integrates the effects of possible geometric imperfections in the structure.

10.1.4. *Mastering cracking*

The Eurocode proposes calculating the crack opening. This concept is new; it was not present in the BAEL regulations and was replaced by the concept of tensile stress on steel but appeared in previous calculation codes (CCBA 68, BA 60, etc.).

A section is assumed not to crack if the tensile stress of the concrete by flexure does not exceed f_{ctm} and if a minimum reinforcement is provided. The national appendices set the maximum values for crack openings according to the exposure classes of the structures. Table 10.8 gives the values defined in the French AN.

Notably:

– the 0.3 mm value corresponds to harmful cracking per the BAEL regulations;

– the value of 0.2 mm corresponds to very damaging cracking per the BAEL regulations.

Table 1a Limits of openings permitted by Eurocode 2

Exposure classes (EN 206-1)	Quasi-permanent loads
X0, XC1	0.40 mm(1)
XC2, XC3, XC4 XD1, XD2, XS1, XS2, XS3	0.30 mm
1) For exposure classes X0 and XC1, crack width has no impact on durability and this limit is set to ensure an acceptable appearance. In the absence of any condition regarding appearance, this limit may be less stringent	

Table 1b: amendments made by the National Appendix

Exposure classes	Reinforced concrete elements Quasi-permanent combination of loads
X0, XC1	0.40 mm(1) no fixed limit except at the request of the DPM
XC2, XC3, XC4	0.30 mm(1) (2)
XD1, XD2, XD3, XS1, XS2, XS3	0.20 mm
1) Cracking control is assumed to be ensured by the construction provisions set out in Chapter 9 of Eurocode 2. The Wmass calculation is no longer required. 2) In the case of buildings in use categories A to D (see Eurocode 1), cracking control is assumed to be ensured, except for specific requests in the DPM (Special Contract Documents) by the minimum construction provisions set out in Chapter 9 of Eurocode 2. The Wmass calculation is no longer required	

Table 10.8. *Crack opening limit values*

However, it should be noted that the calculation is carried out for a cover of 2.5–3 cm; for higher cover values, the results are not necessarily the same.

For buildings, EN 1992.1.1 provides for the calculation to be carried out under quasipermanent combination.

For other structures (bridges, tanks, etc.), each text defines the combination of loads.

10.2. The American Code ACI 318 M-19

10.2.1. *Overview*

American normative codes have followed the same developments as French and European codes. The admissible stress method (working stress design (WSD), then allowable stress design (ASD)) was used from the early 1900s until the early 1960s. It was used for the design of liquid-containing reinforced concrete structures because of the low working rate of the materials and therefore the resulting low level of cracking.

The ASD method is authorized for the calculation of metallic structures in the AISC technical specifications (the name of the method has been changed to the allowable strength method). Timber structures (AF PA 2010 code), masonry structures (MSJC 2008 code) and bridges (AASHTO 2012 code) have historically been designed according to the ASD method, but they can also use the limit state method according to the LRFD method (load and resistance factor design). This latter method is identical to the limit states method defined previously. Three limit states are thus defined:

– strengh limit state;

– serviceability limit state;

– fatigue limit state.

This method is described in the ACI 318 code but also in the MSJC 2008. The ACI 318 code sets the minimum requirements for the design and construction of concrete structures. It applies to concrete with a compressive strength greater than 2,500 psi (approximately 17 MPa). For so-called special structures (tanks, silos, blast-resistant structures, earthquake-resistant structures, chimneys, etc.), these regulations are supplemented by specific requirements.

NOTE.– In comparison, EN 1922.1.1 (EC 2) applies to concrete with a characteristic strength greater than 12 MPa (C12/15).

10.2.2. *Calculation requirements*

10.2.2.1. *Basic requirements*

The calculations are carried out using the semiprobabilistic limit states method. The text also considers the possibility of redistributing moments on support (R8.4). The conditions are identical to those of the Eurocodes. The possibility of redistribution depending on the capacity to generate plastic hinges is defined in Figure 10.3.

The actions are combined using the partial coefficient method.

10.2.2.2. *Intended service life*

The intended service life is defined in the different calculation codes; it is generally 50 years or 75 years.

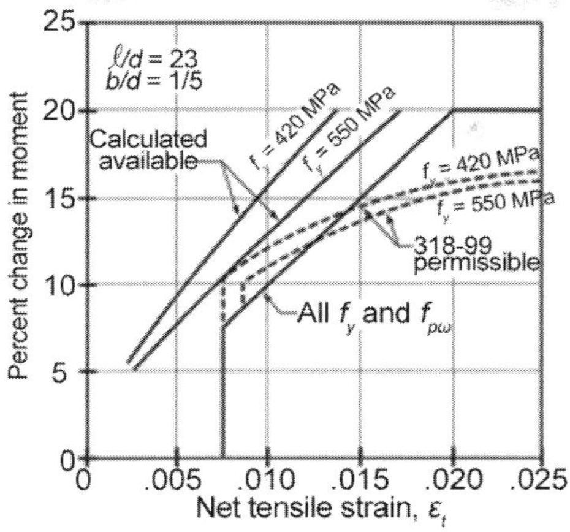

Figure 10.3. *Creation conditions of plastic ball joints*

10.2.2.3. *Thermal effects*

The various loads to be considered are included in the code ASCE/SEI 7.

10.2.2.4. *Settlements and differential movements*

Section 8.2.4 mentions in the comments considering differential settlements.

10.2.3. *Materials*

10.2.3.1. *Concrete*

10.2.3.1.1. Strength

The regulations are applicable to concrete with mechanical resistance characteristics of f'_c = 2,500 psi (17 MPa) and an E/C ratio of approximately 0.45. The compressive strength values are measured at 28 days on cylinders conforming to ASTM C172. The cylinders have dimensions of 4 in diameter by 8 in height or 6 in diameter by 12 in height (10/20 and 15/30, respectively).

The mechanical resistance of concrete under compression is defined as the maximum resistance resulting from the structural calculation and that is necessary for the durability criterion. With respect to the durability criterion, in the case of aggressive environments such as frost or the presence of sulfates, the required values are summarized in Table 10.9 according to the exposure classes selected. The maximum level of chlorides contained in the concrete to ensure the durability of the reinforcements is given in Table 10.13.

10.2.3.1.2. Elastic deformation

The elastic modulus of concrete is given by:

– $E_c = w_c^{1,5}\ 0.043\ f'_c{}^{0.5}$ for concrete density values between 1,440 and 2,560 kg/m^3;

– $E_c = 4{,}700\ f'_c{}^{0.5}$ for standard concrete.

10.2.3.2. *Reinforced concrete steel*

10.2.3.2.1. Properties

The steel used must comply with ASTM A615 (the most commonly used), ASTM A706 (low-alloy steel) and ASTM A996. The elastic limit f_y is generally limited to 60,000 psi (410 MPa). Higher values may be allowed if their elastic limit corresponds to an elongation of 0.35%.

Galvanized steel (conforming to ASTM A707), stainless steel (conforming to ASTM A1022 M) and epoxy resin-covered steel (conforming to ASTM A775) are also permitted. This resin has been used particularly since 1995 in parking lot slabs and bridge decks as well as in structures subject to aggressive environments.

NOTE.– The standard also opens the possibility of using metal fibers for shear resistance only.

10.2.3.2.2. Elastic modulus

The elastic modulus of reinforced concrete steel is taken to be equal to 200,000 MPa.

10.2.4. *Durability and cover of reinforcements*

10.2.4.1. *Overview*

Service life is not explicitly mentioned in the standard. However, Chapter 4 of ACI 318.55 defines the exposure classes of concrete for a duration that can be estimated at 50 years.

10.2.4.2. *Durability requirement*

The requirements of Chapter 4 of ACI 318.55 are presented in Tables 10.9–10.14.

Category	Severity	Class	Condition	
F Freezing and thawing	Not applicable	F0	Concrete not exposed to freezing-and-thawing cycles	
	Moderate	F1	Concrete exposed to freezing-and-thawing cycles and occasional exposure to moisture	
	Severe	F2	Concrete exposed to freezing-and-thawing cycles and in continuous contact with moisture	
	Very severe	F3	Concrete exposed to freezing-and-thawing and in continuous contact with moisture and exposed to deicing chemicals	
S Sulfate			**Water-soluble sulfate (SO_4) in soil, percent by weight**	**Dissolved sulfate (SO_4) in water, ppm**
	Not applicable	S0	$SO_4 < 0.10$	$SO_4 < 150$
	Moderate	S1	$0.10 \leq SO_4 < 0.20$	$150 \leq SO_4 < 1500$ Seawater
	Severe	S2	$0.20 \leq SO_4 \leq 2.00$	$1500 \leq SO_4 \leq 10{,}000$
	Very severe	S3	$SO_4 > 2.00$	$SO_4 > 10{,}000$
P Requiring low permeability	Not applicable	P0	In contact with water where low permeability is not required	
	Required	P1	In contact with water where low permeability is required.	
C Corrosion protection of reinforcement	Not applicable	C0	Concrete dry or protected from moisture	
	Moderate	C1	Concrete exposed to moisture but not to external sources of chlorides	
	Severe	C2	Concrete exposed to moisture and an external source of chlorides from deicing chemicals, salt, brackish water, seawater, or spray from these sources	

Table 10.9. *Table of concrete exposure classes*

Cementitious materials	Maximum percent of total cementitious materials by weight*
Fly ash or other pozzolans conforming to ASTM C618	25
Slag conforming to ASTM C989	50
Silica fume conforming to ASTM C1240	10
Total of fly ash or other pozzolans, slag, and silica fume	50†
Total of fly ash or other pozzolans and silica fume	35†

*The total cementitious material also includes ASTM C150, C595, C845, and C1157 cement.
The maximum percentages above shall include:
(a) Fly ash or other pozzolans in Type IP, blended cement, ASTM C595, or ASTM C1157;
(b) Slag used in the manufacture of an IS blended cement, ASTM C595, or ASTM C1157;
(c) Silica fume, ASTM C1240, present in a blended cement.
†Fly ash or other pozzolans and silica fume shall constitute no more than 25 and 10 percent, respectively, of the total weight of the cementitious materials.

Table 10.10. *Requirements for concrete subject to an exposure class F3*

Exposure Class	Max. w/cm	Min. f'_c, psi	Additional minimum requirements			
			Air content			Limits on cementitious materials
F0	N/A	2500	N/A			N/A
F1	0.45	4500	Table 4.4.1			N/A
F2	0.45	4500	Table 4.4.1			N/A
F3	0.45	4500	Table 4.4.1			Table 4.4.2
			Cementitious materials*—types			Calcium chloride admixture
			ASTM C150	ASTM C595	ASTM C1157	
S0	N/A	2500	No Type restriction	No Type restriction	No Type restriction	No restriction
S1	0.50	4000	II†‡	IP(MS), IS (<70) (MS)	MS	No restriction
S2	0.45	4500	V‡	IP (HS) IS (<70) (HS)	HS	Not permitted
S3	0.45	4500	V + pozzolan or slag§	IP (HS) + pozzolan or slag§ or IS (<70) (HS) + pozzolan or slag§	HS + pozzolan or slag§	Not permitted
P0	N/A	2500	None			
P1	0.50	4000	None			
			Maximum water-soluble chloride ion (Cl^-) content in concrete, percent by weight of cement‖		Related provisions	
			Reinforced concrete	Prestressed concrete		
C0	N/A	2500	1.00	0.06	None	
C1	N/A	2500	0.30	0.06		
C2	0.40	5000	0.15	0.06	7.7.6, 18.16#	

Table 10.11. *Concrete compositions requirements in an aggressive environment*

Nominal maximum aggregate size, in.*	Air content, percent	
	Exposure Class F1	Exposure Classes F2 and F3
3/8	6	7.5
1/2	5.5	7
3/4	5	6
1	4.5	6
1-1/2	4.5	5.5
2†	4	5
3†	3.5	4.5

*See ASTM C33 for tolerance on oversize for various nominal maximum size designations.
†Air contents apply to total mixture. When testing concretes, however, aggregate particles larger than 1-1/2 in. are removed by sieving and air content is measured on the sieved fraction (tolerance on air content as delivered applies to this value). Air content of total mixture is computed from value measured on the sieved fraction passing the 1-1/2 in. sieve in accordance with ASTM C231.

Table 10.12. *Percentage of occluded air as a function of exposure class*

Type of member	Maximum water soluble chloride ion (Cl^-) in concrete, percent by weight of cement
Prestressed concrete	0.06
Reinforced concrete exposed to chloride in service	0.15
Reinforced concrete that will be dry or protected from moisture in service	1.00
Other reinforced concrete construction	0.30

Table 10.13. *Maximum percentage of chloride in the composition of the concrete*

Exposure Class	Maximum expansion when tested using ASTM C1012		
	At 6 months	At 12 months	At 18 months
S1	0.10 percent		
S2	0.05 percent	0.10 percent*	
S3			0.10 percent

*The 12-month expansion limit applies only when the measured expansion exceeds the 6-month maximum expansion limit.

Table 10.14. *Additional requirements for environments containing sulfates*

10.2.4.3. *Cover*

The covers are set at the values (section 7.7) summarized in Tables 10.15–10.17.

Type of concrete	Cover in cm	Comments
Concrete poured in contact with the ground	7.5	
Concrete in contact with the ground or water	Between 4 and 5	The cover depends on the nature of the steel used
Concrete not in contact with the ground or the exterior (slabs, tiles, beams)	Between 2 and 4 cm	The cover depends on the nature of the steel used
Beams, columns	4	

Table 10.15. *Covers for reinforced concrete structures cast in place*

Type of concrete	Cover in cm	Comments
Concrete poured in contact with the ground	7.5	
Concrete in contact with the ground or air: **– slabs, tiles, beams** **– other structures**	2.5 4	
Concrete in contact with the ground but not with the air: **– slabs, tiles, beams** **– Beams, columns**	2 4	The cover for columns and beams depends on the nature of the steel (longitudinal steel or transverse steel, for the latter the cover is 1 in)

Table 10.16. *Covers for prestressed concrete structures cast in place*

NOTE.– In aggressive environments (e.g. attack by chlorides and sea water), section 7.7.6 of the code requires greater coverage. It must not be less than 5 cm for walls and slabs and 6.5 cm for other structures. For the prefabricated elements, these values are reduced to 4 cm and 5 cm, respectively.

Type of concrete	Cover in cm	Comments
Concrete in contact with the ground or air (slabs)	Between 2 and 4	
Concrete in contact with the ground or air (other structures)	Between 3 and 5	The cover depends on the nature of the steel used
Concrete not exposed to contact with air	Between 2 and 3	The cover depends on the nature of the steel used

Table 10.17. *Covers for prefabricated reinforced concrete structures*

10.2.5. *Structural analysis*

10.2.5.1. *Overview*

The purpose of structural analysis is to determine the distributions of stress, strain and deformation. The behavior model used in the American code is that of *linear elastic behavior*. The possibility of redistributing moments on supports is also considered. On the other hand, the plasticity calculation is limited to metal structures.

10.2.5.2. *Load combinations*

The different loads are mentioned in ASCE 7, that is:

– D: live loads;

– Di: weight of ice;

– E: earthquake load;

– F: load due to fluids;

– F_a: flood load;

– H: lateral earth pressure;

– L: live loads;

– L_r: roof live load;

– R: rain load;

– S: snow load;

– T: settlement, differential, temperature, shrinkage and creep (self-straining forces);

– W: wind load.

The combinations of calculations at the SLS are given in section 3 of ASCE 7:

– 1.4D;

– 1.2D + 1.6L + 0.5 (L_r or S or R);

– 1.2D + 1.6 (L_r or S or R) + (L or 0.5W);

– 1.2D + 1.0W + L + 0.5 (L_r or S or R);

– 1.2D + E + L + 0.2S;

– 0.9D + 1.0W;

– 0.9D + E;

– 0.9D + W.

NOTE.– Combinations involving wave and ice loads are described in sections 2.3.3 and 2.3.4.

10.2.5.3. *Second-order effects*

These effects are described in section R10.10 of the ACI Code. Second-order effects can be neglected when the following conditions on slenderness are met:

– lack of bracing: $k\ l_u/r \leq 22$;

– presence of lateral bracing: $k\ l_u/r \leq 34 - 12\ (M_1/M_2)$.

In these formulas, M_1/M_2 is positive if the column is bent in one direction only and negative if it is bent in both directions. The value of factor k is given in Figure 10.4. The length l_u is the length of the compressed zone between the holding points, and r is the radius of gyration.

When it is not possible to disregard second-order effects, the calculation must be carried out either using a nonlinear analysis or second-order linear analysis, taking into account the cracked inertia of the concrete (section R10.10.4). Finally, there is the possibility of amplifying the moments applied to the structure.

CODE COMMENTARY

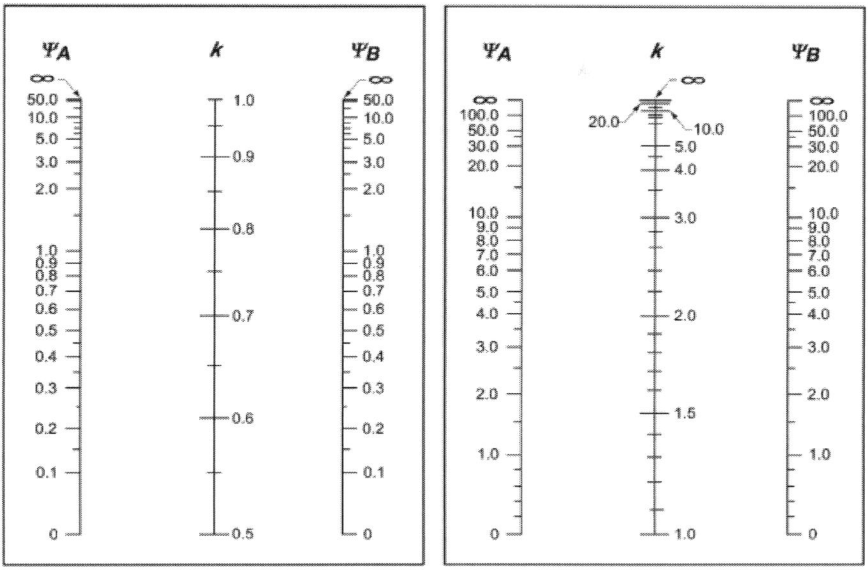

(a) Nonsway Frames

(b) Sway Frames

Ψ = ratio of $\Sigma(EI/\ell_c)$ of compression members to $\Sigma(EI/\ell)$ of flexural members in a plane at one end of a compression member

ℓ = span length of flexural member measured center to center of joints

Figure 10.4. *Values of coefficient k*

10.2.6. *Structural calculations*

10.2.6.1. *Simple compression*

For more details, the reader can refer to section 10.2.6.2.

10.2.6.2. *Compression with bending*

For each of the load combinations defined previously (section 10.2.5.2), the following conditions should be verified:

– $\Phi\ P_n \geq P_u$;

– $\Phi\ M_n \geq M_u$;

– $\Phi\ V_n \geq V_u$;

– $\Phi\ T_n \geq T_u$.

The reduction factor ϕ is given in Table ACI 21.2. Its values are summarized in Table 10.18.

In simple compression, only the first inequality needs to be verified. The value of P_n is then defined according to the nature of the transverse reinforcements:

– for columns with transverse reinforcement as frames: $P_n \leq P_{nmax} = 0.80P_0 = 0.80\,(0.85\,f'_c\,(A_g - A_{st}) + f_yA_{st})$;

– for columns with transverse reinforcement as spirals: $P_n \leq P_{nmax} = 0.85P_0 = 0.85\,(0.85\,f'_c\,(A_g - A_{st}) + f_yA_{st})$.

Net Tensile Strain, ε_t	Classification	Strength Reduction Factor, ϕ — Type of Transverse Reinforcement	
		Spirals Conforming to ACI 25.7.3	Other
$\varepsilon_t \leq \varepsilon_{ty}$	Compression-controlled	0.75	0.65
$\varepsilon_{ty} < \varepsilon_t \leq \varepsilon_{ty} + 0.003$	Transition*	$0.75 + \frac{0.15(\varepsilon_t - \varepsilon_{ty})}{0.003}$	$0.65 + \frac{0.25(\varepsilon_t - \varepsilon_{ty})}{0.003}$
$\varepsilon_t \geq \varepsilon_{ty} + 0.003$	Tension-controlled	0.90	0.90

**Sections classified as transition are permitted to use Ø corresponding to compression-controlled sections.*

Table 10.18. *Values of coefficient ϕ*

These formulas consider a load eccentricity (approximately 10% for columns reinforced with frames and 5% for those reinforced with spirals). The elastic limit of the longitudinal steel f_y is limited to 80,000 psi (544 MPa). P_0 is the resistant force under axial stress without eccentricity.

In compound bending, the following conditions should be verified:

– the maximum relative shortening of the concrete compressed fiber is less than 0.3%;

– the tensile strength of the concrete is neglected;

– the compressive stress of the concrete is limited to $0.85f'_c$;

– the position of the neutral axis is given by the distance a to the fiber compressed by $a = \beta_1\,c$ with:

- $\beta_1 = 0.85$ if 2 500 psi $\leq f'_c \leq$ 4,000 psi,
- $\beta_1 = 0.85 - [0.05\,(f'_c - 4{,}000)/1{,}000]$ if 4,000 psi $\leq f'_c \leq$ 8,000 psi,
- $\beta_1 = 0.65$ if $f'_c \geq$ 8,000 psi.

For a rectangular cross-section, the diagram is shown in Figure 10.5.

The size of the straight cross-section A_g of a column subjected to an axial compression force can be determined as a function of the transverse reinforcement:

– when the transverse reinforcement is composed of frames:

$$A_g \gg \frac{P_u}{\emptyset 0.80\left[0.85f'_c(1-p_g)+f_y p_g\right]}, \emptyset = 0.65$$

– when the transverse reinforcement is composed of spirals:

$$A_g \gg \frac{P_u}{\emptyset 0.85\left[0.85f'_c(1-p_g)+f_y p_g\right]}, \emptyset = 0.75$$

with $\rho_g = A_{st}/A_g$.

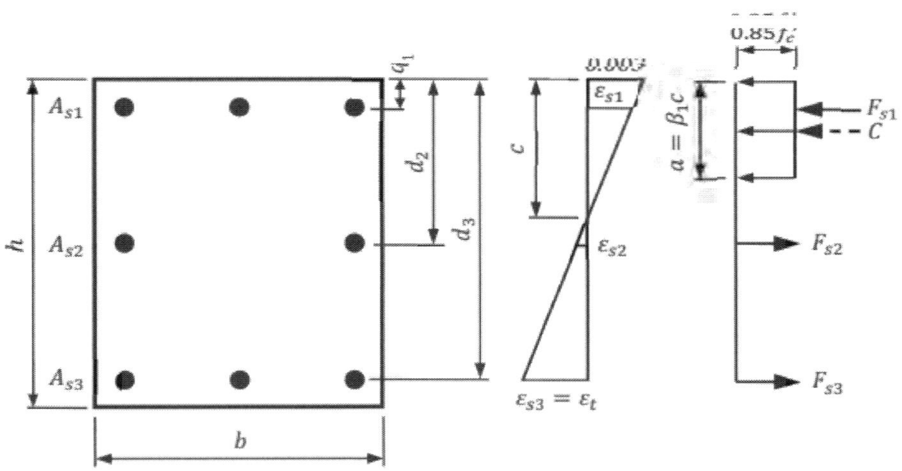

Figure 10.5. *Stress and deformation ratio for a rectangular cross-section*

NOTE.– In predimensioning for a square column, the following values are often considered:

– grade 60 or 80 longitudinal reinforcement;

– compressive strength of concrete: from 4,000 to 14,000 psi;

– longitudinal reinforcement ratio of 1–2%.

The dimensioning of the longitudinal reinforcement A_{st} required under axial compression (without taking into account the effects of slenderness) is given by:

$$p_g = \frac{\left(\frac{P_u}{\varnothing\varnothing_c A_g}\right) - 0.85f'_c}{f_y - 0.85f'_c}$$

$$\gg P_{g,min} = 0.01$$

$$\ll P_{g,max} = 0.08$$

In this equation:

– $\rho_g = A_{st}/A_g$;

– ϕ is 0.65 when the transverse reinforcement consists of frames and 0.75 for spirals;

– ϕ_c is 0.80 when the transverse reinforcement consists of frames and 0.85 for spirals.

In the case of compound bending, the reinforcement is calculated using bending moment–normal force interaction diagrams. Appendix C provides access to diagrams for rectangular and circular columns.

The transverse reinforcement A_v is given as a function of the spacing s of the steel according to the following formula:

$$s = \frac{\varnothing A_v f_{yt} d}{V_u - \varnothing V_c}$$

The minimum longitudinal reinforcement of the columns is:

– 3 when the transverse reinforcements are triangular;

– 4 when the transverse reinforcements are frames or hoops;

– 6 when the transverse reinforcements are spirals.

The spacing of the longitudinal steel is the maximum value (1.5 in; 1.5 d_b; 4/3 d_{agg}), where d_b is the diameter of the longitudinal bars and d_{agg} is the diameter of the largest aggregate.

The continuity of longitudinal reinforcement is given in section ACI 10.7.5. This can be achieved by overlapping, welding or couplers.

10.2.6.3. *Bending*

Beam support conditions	h/l for f_y = 60 kpsi	h/l for f_y if not 60 kpsi
Simple supports	l/16	(l/16) f_l
Continuous support	l/18.5	(l/18.5) f_l
Two continuous supports	l/21	(l/21) f_l
Cantilever	l/8	(l/8) f_l

Table 10.19. *Maximum flexure values*

The size of the bending section is determined by the flexure condition and summarized in Table 10.19 for normal concrete (with $f_l = 0.4 + (f_y/100{,}000)$, h beam height and l span).

The width of the beam b_w is generally given in predimension and considers a minimum cross-section of reinforcements of $0.01b_wd$ by the following formula:

$$b_w = \frac{M_u}{0.009d^2 f_y(1 - \frac{0.0059f_y}{f'_c})}$$

ACI Code 6.2.3 defines five analysis methods for the design of bending moments and shear forces. The simplified method proposed in ACI section 6.5 can be used for beams:

– prisms;

– subjected to uniformly distributed gravitational loads;

– whose operating load is less than three times the live load;

– comprising at least two lengths;

– whose ratio between the spans of adjacent spans is less than 1.20.

For a cross-section subjected to bending tension, the value of the longitudinal reinforcement is given by the following formula:

$$A_s = \frac{0.85f'_c bd}{f_y}[1 - \sqrt{1 - \frac{2R_n}{0.85f'_c}}]$$

where $R_u = M_u/(\phi\ b\ d^2)$ and $\phi = 0.90$.

The minimum reinforcement is A_{Smin}, such that $A_{smin} \geq$ Max ((3 b_w d $(f'_c)^{0.5}$)/fy; 200 b_w d/f_y). For a rectangular cross-section where traction is not controlled, compressed reinforcement can add A'_s according to the following formula:

$$A'_s = \frac{M'_n}{f'_s(d - d')}$$

The total section of the steel then becomes:

$$A_s = pdb + \frac{M'_n}{f_y(d - d')}$$

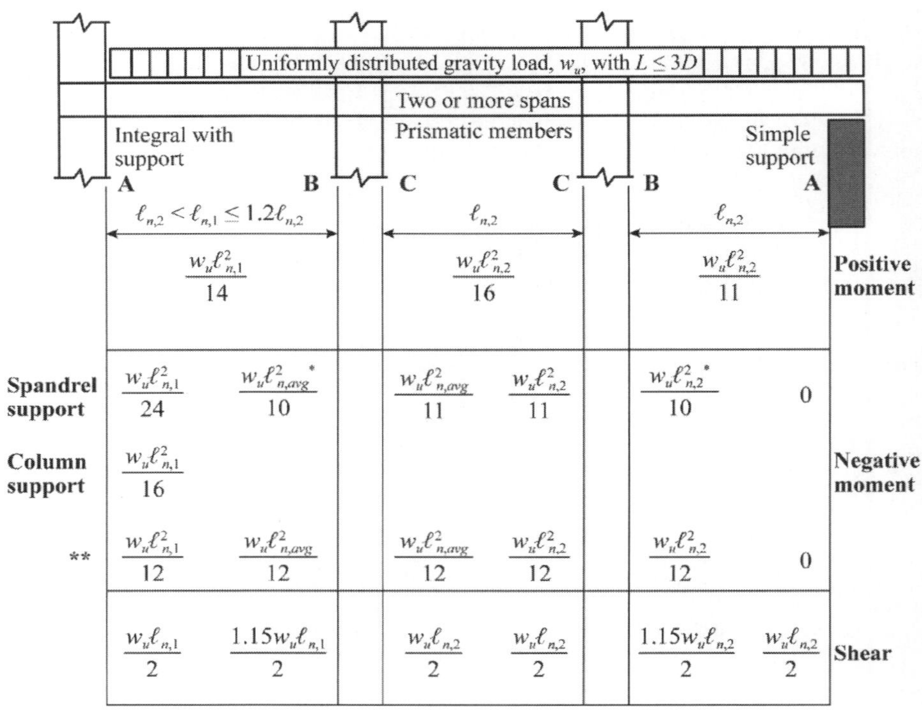

Table 10.20. *Simplified values of bending moments according to the CRSI*

10.2.7. *Constructive provisions*

10.2.7.1. *Concrete pouring*

The concrete implementation is defined in the ACI 304R.00 code. The drop height is limited to 24″ (0.60 m). Section 5.4 highlights the need to provide pouring and vibrational stacks at the reinforcement level. The vibration methods are described in the ACI 309R code. For large mass castings, it is recommended not to exceed layer heights of 610 mm.

10.2.7.2. *Distance between reinforcements and between reinforcements and formwork*

The minimum values of the covers are given in ACI 20.5.1.

– For slabs, the distance between reinforcements must be greater than Max (1 in; d; (4/3) d_{agg}). In this formula, d_b denotes the diameter of the bar and d_{agg} represents the diameter of the largest aggregate.

The maximum spacing s of the steel is given by: $s \leq$ Min (15 (40,000/f_s) – 2.5 c_c; 12 (40,000/f_s)). c_c represents the distance between the surface of the reinforcement and the tensioned face of the concrete; f_s denotes the tensile stress of the steel.

– For beams, the distance between reinforcements must be greater than Max (1 in; d_b; (4/3) d_{agg}). In this formula, d_b denotes the diameter of the bar and d_{agg} represents the diameter of the largest aggregate.

The maximum spacing s of the steels is given by: $s \leq$ Min (15 (40,000/f_s) – 2.5 c; 12 (40,000/f_s)). c_c represents the distance between the surface of the reinforcement and the tensioned face of the concrete, and f_s denotes the tensile stress of the steel.

– For columns, the spacing of longitudinal steel is the maximum value (1.5 in; 1.5 d_b; 4/3 d_{agg}), where d_b is the diameter of the longitudinal bars and d_{agg} is the diameter of the largest aggregate.

In the bar overlap areas, the following provisions are specified (Figure 10.6).

For columns, the minimum frame cover is 1.5 for structures not exposed to weather or in contact with the ground.

– For slabs, the maximum spacing s of the longitudinal steel is given by: $s \leq$ Min (3 h; 18 in; l_w/3). h refers to the slab thickness, and l_w is the slab length. The maximum spacing s of the transverse steel is given by: $s \leq$ Min (3 h; 18 in; l_w/5). This latter condition is only applicable when reinforcement is necessary to absorb shear stresses.

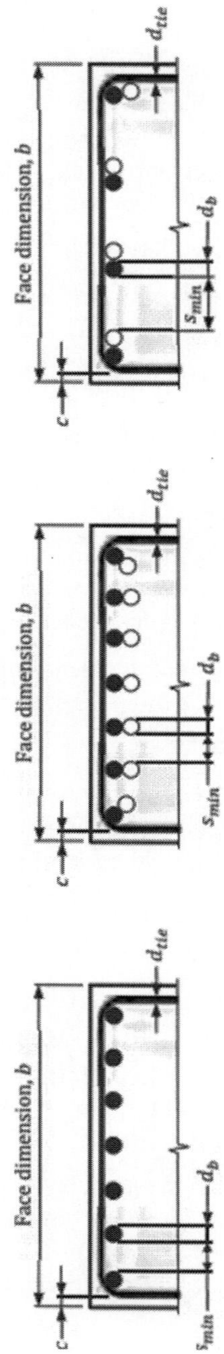

Figure 10.6. *Construction details in the overlap areas*

10.2.7.3. *Concrete joints*

Chapter 6.2 specifies the method for processing concrete joints. Therefore, the repair surfaces must be cleaned. In most cases, this involves removing the slurry and hydrocleaning or hydrosanding the surface. Conventional hand tools can also be used.

10.2.7.4. *Concrete curing*

The concrete curing conditions are given in ACI 308.

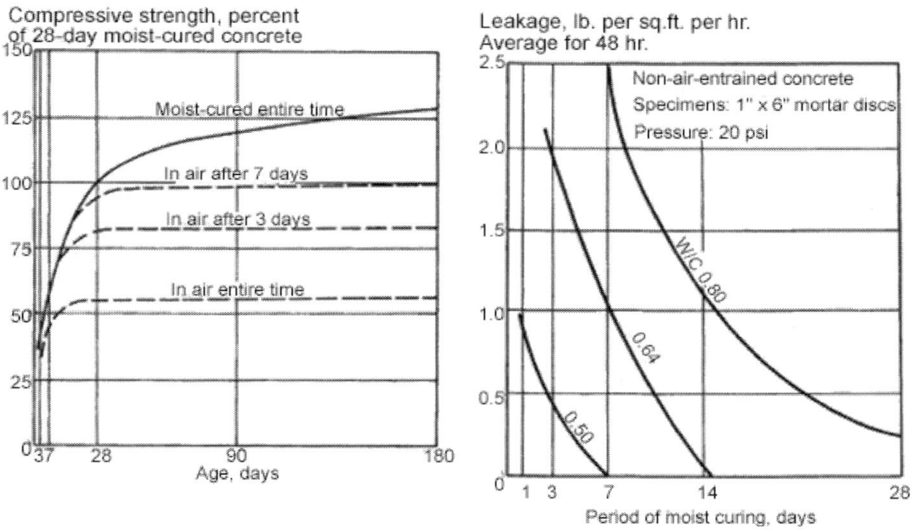

a) b)

Figure 10.7. *(a) Compressive strength values of concrete with a curing product as a function of time. (b) Influence of curing on the permeability of concrete*

The code specifies the different methods that can be used for curing concrete:

– use of water: deposit a homogeneous film of water on the surface of the concrete;

– use of water mist;

– use of curing products;

– use of moistened materials (cotton, absorbent materials, straw, etc.);

– use of plastic film.

The processing times are listed in Table 10.21.

	Minimum curing period
ASTM C 150 type I	7 days
ASTM C 150 type II	10 days
ASTM C 150 type III or when accelerators are used to achieve results demonstrated by test to be comparable to those achieved using ASTM C 150 type III cement	3 days
ASTM C 150 type IV or type V cement	14 days
Blended cement, combinations of cement and other cementitious materials of various types in various proportions in accordance with ASTM C 595, C 845 and C 1157	Variable. See section 2.9

Table 10.21. *Curing time depending on the concrete composition. With various cement types when no testing is performed and no concrete properties are specified*

10.3. The Russian standard SP63.13330.2018

10.3.1. *Overview*

This set of regulations applies to the design of concrete and reinforced concrete structures and for various purposes, specifically for the climatic conditions of the Russian Federation (with systematic exposure to temperatures not exceeding 50°C and not lower than –70°C) in a non-aggressive environment. The typical service life of ordinary buildings is 50 years.

10.3.2. *Calculation bases*

In the calculation of elements under loads related to handling in the provisional phase, lifting and transport, the live weight of the elements must be multiplied by the dynamic coefficient equal to:

– 1.6 in the case of transport loads;

– 1.4 in the case of handling and lifting loads.

It is permissible to take into account smaller values, if duly justified. This dynamic coefficient should never be less than 1.25.

Concrete or reinforced concrete elements must be justified at the limit states:

– First group (ULS, causing total unsuitability for the destination, equivalent to ULS). The shape stability, mechanical resistance, and position stability are calculated.

– Second group (SLS, which causes difficulties in use at destinations or reduces the durability of buildings and constructions, equivalent to the SLS). The crack formation, crack opening, and deformation were calculated.

10.3.3. *Requirements*

The basic requirements of the standard are as follows:

– requirements for mechanical strength calculation: $F \leq F_{ult}$;

– requirements for the calculation of crack formation: $F \leq F_{crc,ult}$:

– requirements for calculating crack openings: $a_{crc} \leq a_{crc,ult}$.

$a_{crc,ult}$	**Bar strength classes**	**Reinforcement protection condition**
0.3 mm	A240-A600, B500	Long-lasting crack opening
0.4 mm		Short-term crack opening
0.2 mm	A800, A1000, Bp1200-Bp1400, K1400, K1500 (K-19), K1500 (K-7), K1600 diameter 12 mm and more	Long-lasting crack opening
0.3 mm		Short-term crack opening
0.1 mm	Bp1500, K1500 (K-7, K1600-K1900 of diameter less than 12 mm)	Long-lasting crack opening
0.2 mm		Short-term crack opening
		Waterproofing condition of the elements
0.2 mm		Long-lasting crack opening
0.3 mm		Short-term crack opening

Table 10.22. *Limit values for permissible crack width*

Requirements for calculating deformations are as follows: $f \leq f_{ult}$.

10.3.4. *Project life time*

The duration of use is set at 50 years for a project in the normal importance category and 100 years for a project of high importance or belonging to heritage.

10.3.5. *Thermal effects*

The standard provides for the construction of thermal expansion joints according to Russian climatic conditions, that is:

– every 50 m for heated buildings;

– every 40 m for unheated buildings;

– every 30 m for structures open to the outside air.

10.3.6. *Settlements and differential movements*

The standard does not provide information on the subject.

10.3.7. *Materials*

10.3.7.1. *Concrete*

10.3.7.1.1. Strength

For reinforced concrete elements, the strength class must be at least B15 (15 MPa). For prestressed concrete elements, the strength class must be at least B20 (20 MPa).

The compressive strength values are measured at 28 days on cubes with dimensions of 15 × 15 × 15 cm (GOST 18105-2018). The value of the design axial compressive strength is: $R_b = R_{b,n}/\gamma_b$, where $R_{b,n}$ denotes the base resistance value and γ_b is the safety coefficient of the material, which has a value of:

– 1.3 for ordinary concrete with density between 2,200 and 2,500 kg/cm;

– $\gamma_{b2} = 0.9$ for unreinforced concrete elements;

– $\gamma_{b3} = 0.85$ for vertical elements in unreinforced and reinforced concrete; the layer height when concreting above 1.5 m;

– $\gamma_{b5} \leq 1$ for items subjected to freezing/thawing or negative temperatures.

The concrete strength must be multiplied by a coefficient of the concrete working condition.

Regarding the durability criterion, in the case of aggressive environments such as frost or the presence of aggressive agents, the required values are summarized in Tables 6.3 and 6.4 of standard SP63.13330, and the choice will be made according to the exposure conditions retained according to standard SP28.13330 "Protection of construction elements against corrosion".

10.3.7.1.2. Elastic deformation

The Young's modulus and design values of the concrete compressive strength can be found in Table 10.23.

Concrete class	**B15**	**B20**	**B25**	**B30**	**B35**	**B40**
Young's modulus ($MPa/10^3$)	24	27.5	30	32	34.5	36
Compressive strength R_b (MPa)	8.5	11.5	14.5	17	19.25	22

Table 10.23. *Value of Young's modulus in common concrete*

The unit elongation is limited to:

– $\varepsilon_{b0} = 0.002$ for axial compression;

– $\varepsilon_{bt0} = 0.0001$ for axial traction.

The creep coefficient can be found in Table 10.24 (some values for a humidity level greater than 75%).

Concrete quality	**B15**	**B20**	**B25**	**B30**	**B35**	**B40**
Creep coefficient values	2.4	2	1.8	1.6	1.5	1.4

Table 10.24. *Creep coefficient values*

The stress–deformation diagram of the concrete is a bilinear diagram according to Figure 10.8.

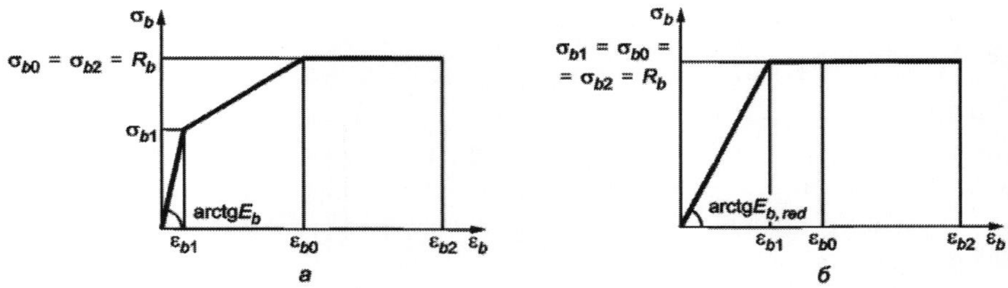

Figure 10.8. *Stress–deformation diagram of the concrete*

10.3.7.2. *Reinforced concrete steel*

The main reinforcement characteristic is the tensile strength (class A for cold-formed and thermally hardened bars, B for cold-formed bars). The classes correspond to the normative tensile strength of the bars $R_{s,n}$, (MPa). Certain classes of reinforcement cannot be used in elements located in open air at temperatures lower than –30°C (the standard specifies steel with its use limits).

10.3.7.2.1. Properties

For non-prestressed reinforced concrete, the main bars are of class A400, A500, A600 and B500 in cages and meshes. The secondary bars are most often of class A240. For prestressed reinforced concrete, the prestressed bars can be of the following classes: A600, A800, A1000, Bp1200-Bp1600 and K1400-K1900 (cables).

10.3.7.2.2. Strength

The reinforcement classes correspond to the standard tensile strength of the bars $R_{s,n}$ (MPa). The design resistance will be equal to $R_s = R_{s,n}/\gamma_s$, where $\gamma_s = 1.15$ for the ULS and 1 for the SLS (the safety coefficient of the reinforcements).

The design compressive strength is equal to the tensile strength but is never greater than:

– 400 MPa in the case of short-term loads;

– 500 MPa in case of live loads.

The elastic modulus of steel is equal to:

– $Es = 2 \times 10^5$ MPa for the bars of resistance classes A and B;

– $Es = 1.95 \times 10^5$ MPa for resistance class K bars (cables).

The characteristic diagrams of steel are bilinear diagrams (Figure 10.9).

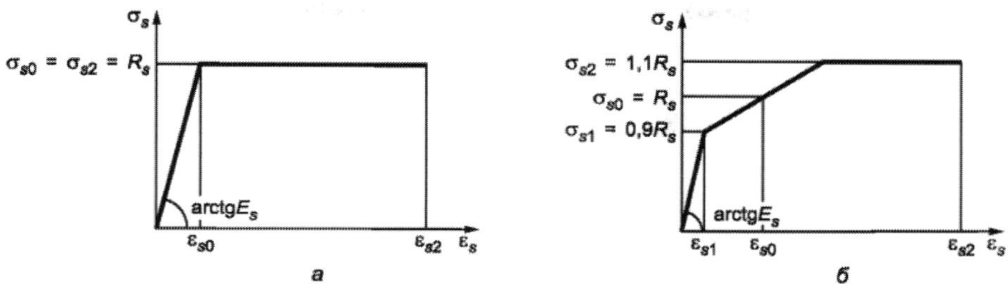

Figure 10.9. *Stress–deformation diagrams of steel*

10.3.7.2.3. Steel/concrete adhesion

The bars can be profiled or smooth. The different types of anchoring of reinforcements in concrete are given by the standard:

– straight anchor (straight end of a bar in concrete for bars with a profiled surface only);

– anchoring with bending or hook (except reinforcements for prestressed concrete);

– anchoring with the welding of additional crossbars (except reinforcements for prestressed concrete);

– installation of U-shaped anchor reinforcements or welded mesh in the compressed zone of the concrete (except reinforcement for prestressed concrete);

– anchoring with special elements at the end of the bars to be anchored (plates, washers, nuts, etc.).

10.3.8. *Durability and concrete cover*

For more details, please refer to standard SP28.13330.2017: protection of construction elements against corrosion.

10.3.8.1. *Overview*

The lifespan of ordinary buildings is 50 years.

10.3.8.2. *Durability requirement*

Some requirements of the appendices to standard SP28.13330.2017 are recalled in Table 10.25.

Definition	Associated environment	Example of building elements
X0	No aggressiveness, dry environment for concrete	Elements located indoors, in rooms with low humidity
Corrosion of reinforcements due to carbonation		
XC1	Constantly dry or humid environment	Elements inside residential buildings, except for very humid rooms, such as kitchens, bathrooms, dry cleaners, laundries, etc.
XC2	Humid environment, occasionally dry	Concrete surfaces permanently moistened. Foundations
XC3	Medium humidity environment (humid premises, humid climate)	– Building elements, often or constantly exposed to the outside air, without being humidified by atmospheric precipitation – Elements located under a canopy – Elements located in rooms with high humidity (communal kitchens, bathrooms, laundries, indoor swimming pools, livestock premises)
XC4	Alternate drying/humidification	Elements located outside, under the action of atmospheric precipitation
Corrosion as a result of the action of chlorides (except sea water)		
XD1	Moderate humidity environment	Elements under splashes of chloride salts
XD2	Humid environment, occasionally dry	Swimming pools, elements subject to the action of industrial wastewater, containing chlorides
XD3	Alternate drying/humidification	– Bridge elements exposed to de-icing salts – Road surfaces – Parking lot floors
Corrosion under the action of sea water		
XS1	Splashing, but without direct contact with sea water	Dock elements
XS2	Underwater	Submerged elements
XS3	Tidal zone, splashing	Parts of the elements subject to humidification according to the tides
Chemical or biological aggression		
XA1	Low quantity of aggressive agents (see corresponding tables)	Elements bathed in the waters of the aquifer
XA2	Average quantity of aggressive agents (see corresponding tables)	– Elements in contact with seawater – Elements in aggressive soils
XA3	High amount of aggressive agents (see corresponding tables)	– Water treatment plant components – Livestock feeders – Cooling towers

Table 10.25. *Concrete exposure conditions*

10.3.8.3. *Cover*

The covers are set to the values in Table 10.26 with a minimum of 10 mm or equal to the diameter of the bars. For secondary bars, it is permissible to reduce the values in the table to 5 mm.

– Reinforced concrete cast in place (see Table 10.26).

Type of concrete	Cover in mm
Concrete poured in contact with the ground, or foundations with blind concrete	40
Concrete in contact with outside air	30
Concrete in contact with indoor air in humid rooms	25
Concrete in contact with indoor air in moderately humid and dry rooms	20

Table 10.26. *Steel cover*

– *Prefabricated reinforced concrete elements*: the values retained are those of Table 10.26 minus 5 mm.

10.3.9. *Structural analysis*

10.3.9.1. *Overview*

The principle is that of limit states. At the ULS, the resistance of reinforced concrete elements is verified under the stresses of the bending moment, shear force, axial force, torsional torque and local load (local compression and punching). The behavior model used is that of nonlinear behavior and is based on the notion of limit forces (that the element is capable of supporting).

The limit forces are defined as follows:

– bending moment $M \leq M_{ult}$;

– compression with eccentricity $N \leq N_{ult}$;

– axial traction $N \leq N_{ult}$.

Nonlinear behavior model: In this model, the equations of equilibrium of the external forces and internal forces in a section of an element are used. In addition, the following hypotheses are considered:

– the deformation distribution in the concrete and reinforcement over the height of the section is linear;

– the relationship between axial stresses and deformations in the concrete and reinforcement is represented by stress–deformation diagrams;

– the concrete strength in the tensioned zone is neglected (in certain special cases, it can be taken into account, for example, for elements bent or compressed with eccentricity, in the absence of cracks requirement).

An accidental eccentricity is also defined, the value of which is the maximum of the following values:

– 1/600 of the length of the element or the distance between its fixed points;

– 1/30 of the cross-section height;

– 10 mm.

10.3.9.2. *Load combinations (excerpts from SP20.13330.2016 loads and influences)*

The load design value is equal to the normative load multiplied by the load safety coefficient γ_f. This coefficient is equal to 1 for the SLS. The design loads for wind, snow, ice and thermal gradient can be reduced by 20% for temporary phases (construction period). The ice load must be considered for power lines and connection lines, transport power lines, high structures in the form of masts, fencing, antennas, ventilated building facades, balcony railings, walls and roofing located higher than 100 m.

Load classification:

– live loads P_d (live weight, ground pressure and hydrostatic pressure);

– variable charges: P_l long term (weight of partitions, leveling compounds and screeds, weight of equipment and weight of storage and racks), P_t short term (human and animal loads on floors, rolling lifting devices and overhead crane loads) and P_s specific loads (earthquake loads, accidental loads, such as explosion, equipment failure, ground deformation loads with change in the structure of the supporting soils, loads following a fire, transport accidents).

Load	Description	γ_f
Self-weight of structures	Metal structures	1.05
	Concrete structures (with unit weight > 1,600 kg/m³), reinforced concrete, masonry, wood	1.1
	Concrete structures (with a unit weight ≤ 1,600 kg/m³), insulation materials, screeds, leveling compounds, installed: – in the factory – on site	1.2 1.3
Self-weight of soil	In a natural layer On the construction site	1.1 1.15
Equipment and materials	Fixed equipment	1.05
	Insulation of fixed equipment	1.2
	Forklifts, lifting machines (with load)	1.2
	Stored materials and products	1.2
Live loads	Normative load value less than 2 kPa (200 kg/m²) Normative load value equal to or greater than 2 kPa (200 kg/m²)	1.3 1.2
Vehicle loads	Parking lots	1.2
Snow load		1.4
Wind load		1.4
Ice load		1.8
Temperature gradient		1.1

Table 10.27. *Values of the coefficient γ_f*

Load combinations: There are two possible combinations:

1) main combination (live loads + long-term variable loads + short-term variable loads):

$$C_m = P_d + (1 \times P_{l1} + 0.95 \times P_{l2} + 0.95 \times P_{l3} + \ldots + 0.95 \times P_{li}) + (1 \times P_{t1} + 0.9 \times P_{t2} + 0.7 \times P_{t3} + 0.7 \times P_{t4} + \ldots + 0.7 \times P_{ti})$$

2) specific combination (live loads + long-term variable loads + short-term variable loads + one specific load):

$$C_s = C_m + P_s$$

where:

$$C_m = P_d + (1 \times P_{l1} + 0.95 \times P_{l2} + 0.95 \times P_{l3} + \ldots + 0.95 \times P_{li}) + (0.5 \times P_{t1} + 0.3 \times P_{t2} + 0.3 \times P_{t3} + 0.3 \times P_{t4} + \ldots + 0.3 \times P_{ti})$$

Load safety coefficient γ_f: the values of γ_f are summarized in Table 10.27.

10.3.9.3. *Second-order effects*

The standard does not provide any guidance on the subject.

10.3.9.4. *Structural modeling*

Modeling of construction systems can be carried out using finite elements, bars, plates and volume elements (Appendix B). When creating a model of a three-dimensional system, it is necessary to consider the simultaneous working mode of the elements (bars, plates and solids). Each type of element can have a different degree of freedom.

As a first approach and in the absence of theoretical reinforcement data, the stiffnesses of the elements are defined based on their linear deformation behavior. After the necessary reinforcement sections in the slabs are defined, the deflections of these elements should be recalculated, taking into account the precise stiffness characteristics of the elements, with the reinforcement data in both directions. It is recommended to carry out an additional calculation of the construction scheme, taking into account the nonlinear deformation behavior in the elements, to specify the values of the bending moments in the slabs and rafts and the axial forces in the walls and columns.

For important buildings of class KSC-3 (high importance, GOST 27751), the size must be calculated using two different software programs by two independent engineering firms. The documentation must be reviewed by an independent third party.

10.3.10. *Structural calculations*

10.3.10.1. *Compression with eccentricity*

Accidental eccentricity e_a is given by the largest of the following values:

– 1/600 of the length of the element or the distance between its fixed points;

– 1/30 of the section height;

– 10 mm.

The calculation formula is as follows:

$$N \cdot e \leq R_b \cdot b \cdot x\left(h_0 - 0.5x\right) + R_{sc} \cdot A_s^{'}\left(h_0 - a'\right)$$

where:

– N is the axial force;

– e the distance between the point of application of force N and the center of gravity of the reinforcement bar cross-section, given by the following formula:

$$e = e_0 \cdot \eta + \frac{h_0 - a'}{2}$$

The value of the coefficient η is as follows:

$$\eta = \frac{1}{1 - \dfrac{N}{N_{cr}}}$$

The critical Euler force is then equal to:

$$N_{cr} = \frac{\pi^2 \cdot D}{l_0^2}$$

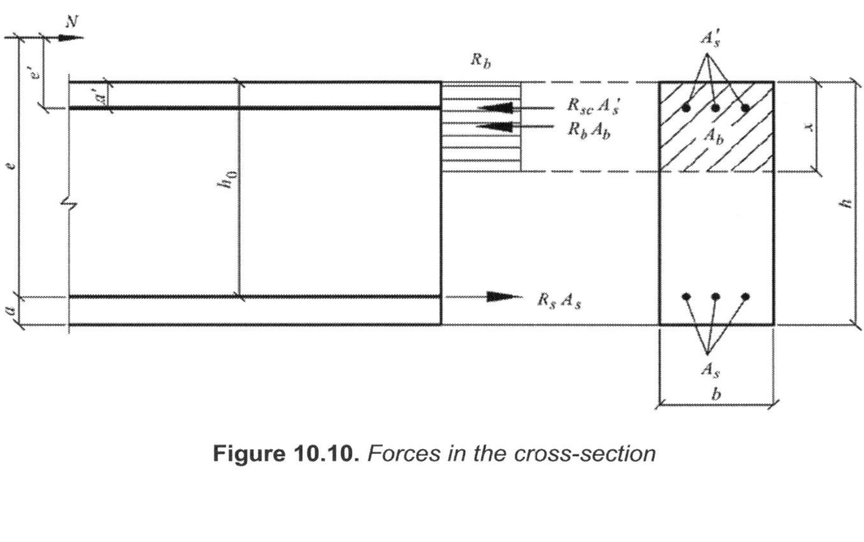

Figure 10.10. *Forces in the cross-section*

The cross-sectional functions are summarized in Figure 10.10.

The design length of the elements depends on the support conditions according to Figure 10.11.

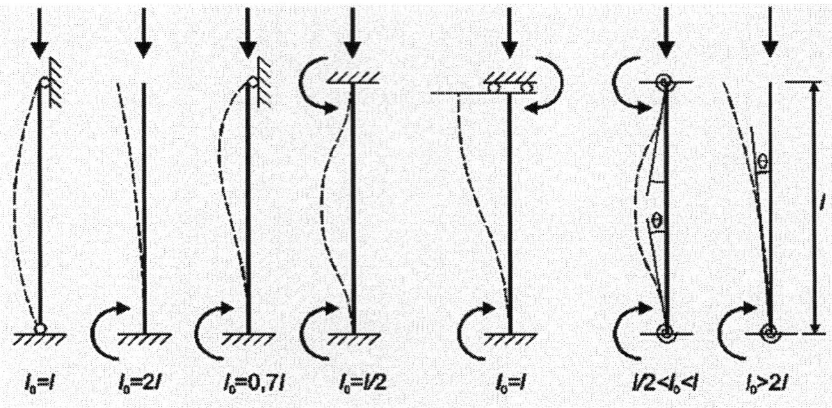

Figure 10.11. *Value of buckling lengths*

10.3.10.2. *Compression with flexion*

The standard does not provide specifications for the subject. For more details, see section 10.3.10.3.

10.3.10.3. *Bending*

The standard specifies the conditions to be respected for rectangular cross-sections: $M \leq M_{ult}$. The strength of the normal cross-sections is calculated if the height of the compressed zone of the concrete is:

$$\xi = \frac{x}{h_0} \leq \xi_R$$

is:

$$M_{ult} = R_b \cdot b \cdot x \cdot (h_0 - 0.5x) + R_{sc} \cdot A_s' \cdot (h_0 - a')$$

Otherwise, if:

$$\xi > \xi_R$$

then:

$$M < \alpha_R R_b b h_0^2 + R_{sc} A_s' (h_0 - a')$$

with:

$$\alpha_R = \xi_R (1 - 0.5\xi_R)$$

$$x = \frac{R_s \cdot A_s - R_{sc} \cdot A_s'}{R_b \cdot b}$$

The values of α_R and ξ_R are given according to the steel strength class in Table 10.28.

Класс арматуры	A240	A400	A500	A500СП	A600	Ан600С	B500
Значение ξ_R	0.615	0.533	0.493	0.487	0.459	0.337	0.493
Значение α_R	0.426	0.391	0.372	0.368	0.354	0.280	0.372

Table 10.28. *Values of α_R and ξ_R depending on the grade of steel*

The operating diagram of the cross-section is identical to Figure 10.12.

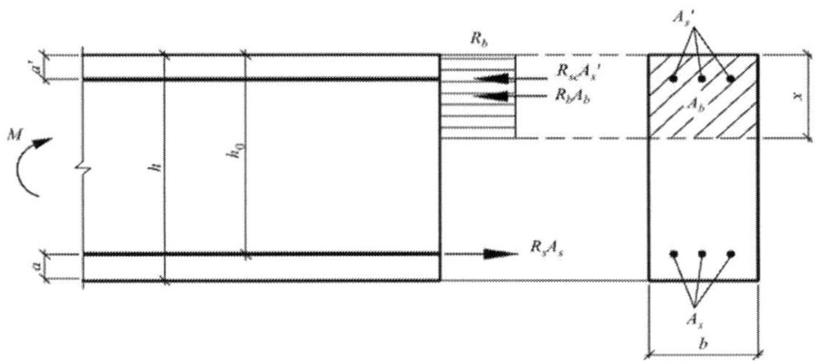

Figure 10.12. *The stress and deformation diagram in the cross-section of the bent element*

10.3.10.4. *Buckling*

The standard does not provide any guidance on this topic.

10.3.11. *Constructive provisions: standard SP70.13330.2012*

10.3.11.1. *Pouring concrete*

Concrete pumps or backfillers are permitted for pouring concrete, with a minimum intensity of 6 m^3/h. The maximum concreting height is given in Table 10.29.

Type of structure	Pouring height (m)
Posts	3.5
Slabs	1
Walls	4.5
Structures without reinforcement	6
Underground lightly reinforced structures	4.5
Heavily reinforced structures	3

Table 10.29. *Maximum concrete pouring height*

10.3.11.2. *Distance between reinforcements and between reinforcements and formwork*

The distance between reinforcements must be at least equal to:

– the largest diameter of the bars;

– 25 mm for horizontal or inclined bars;

– 50 mm for vertical bars.

The maximum distance between bars is:

– for slabs and beams, 200 mm if $h \leq 150$ mm, otherwise 1.5h or 400 mm;

– for columns, 400 mm in the direction perpendicular to the bending plane; otherwise, 500 mm;

– for walls, 2t or 400 mm for vertical bars and 400 mm for horizontal bars;

– for transverse reinforcements, $0.5h_0$ or 300 mm.

10.3.11.3. *Treatment of the construction joint*

The construction joint should be perpendicular to the axis of the element. The joint is authorized once the concrete has reached a strength of 1.5 MPa. The position of the concrete joints must be validated by the engineering design office.

10.3.11.4. *Concrete curing*

The surface of fresh concrete must be protected against water evaporation and atmospheric precipitation. A state of optimal temperature and humidity must be maintained (heating in winter with a heat gun or electric cables or watering/humidification) until 70% strength is achieved.

10.3.11.5. *Concrete control tests*

They are given in section 11.5 for testing concrete samples (cubes) or non-destructive testing (GOST 18105-2018). Two batches of samples must be taken from each concrete batch.

10.3.11.6. *Constructive arrangements for the reinforcement of the columns*

The columns are reinforced by longitudinal bars at the contour of the section and inside the cross-section if justified by the calculation and by transverse reinforcements at the contour placed around the vertical bars.

Appendices

Appendix 1

Characteristics of the Steel Used in the Different Regulations

	High-grip steel
Designation	TS
Nominal diameters (mm)	3; 3.5; 4; 4.5; 5; 5.5; 6; 7; 8; 9; 10; 12
Wire spacing	Carrier wires: 75; 100; 125; 150; 200 Distribution wires: 50; 100; 150; 200; 250; 300
Elastic limit (bar)	$\Phi \leq 6$ mm 5,200 $\Phi \geq 6$ mm 4,500
Breaking stress (bar)	$\Phi \leq 6$ mm 6,200 $\Phi \geq 6$ mm 5,150
Elongation at break (%)	Not guaranteed
Elastic limit taken into account in the calculations (bar)	$\Phi \leq 6$ mm 5,200 $\Phi \geq 6$ mm 4,250
Cracking coefficient	1

Table A1.1. *Mechanical and geometric characteristics of welded mesh*

Mild steel	Mild steel	Steel A42	Steel A55	Steel AC54	Isteg steel	Steel A65	Steel A75
Designation	Adx	A42	A55	AC54	Isteg	A65	A75
Nominal diameters (mm)	5, 6, 8, 10, 12, 16, 20, 25, 32, 40	5, 6, 8, 10, 12, 16, 20, 25, 32, 40	5, 6, 8, 10, 12, 16, 20, 25, 32, 40	5, 6, 8, 10, 12, 16, 20, 25, 32, 40	Undefined	5, 6, 8, 10, 12, 16, 20, 25, 32, 40	5, 6, 8, 10, 12, 16, 20, 25, 32, 40
Elastic limit (bar)	2,400	2,400	3,200	3,600	3,600	3,600	4,000
Breaking stress (bar)	3,300–5,000	4,200–5,000	5,500–6,500	5,400–6,400	4,000	6,500–8,500	≥7,500
Elongation at break (%)	25	25	16	20	10	12	6
Elastic limit taken into account in the calculations (bar)	2,400	2,400	3,200	3,600	3,600	3,600	4,000

Table A1.2. *Summary table of high-grip steel characteristics used according to the various regulations*

High grip steel	Tentor steel	Tor steel	Caron steel	Nersid steel	Créloi steel
Designation	TT	T	C	N	SE
Nominal diameters (mm)	6, 8, 10, 12, 14, 16, 20, 25, 32	5, 6, 8, 10, 12, 14, 16, 20, 25, 32, 40	5, 6, 8, 10, 12, 14, 16, 20, 25, 32	8, 10, 12, 14, 16, 20, 25, 32, 40	8, 10, 12, 14, 16, 20, 25
Elastic limit (bar)	5,000	$\Phi \leq 20$ mm 4,200 $\Phi \geq 25$ mm 4,000	$\Phi \leq 20$ mm 4,200 $\Phi \geq 25$ mm 4,000	$\Phi \leq 16$ mm 4,200 $\Phi \geq 20$ mm 4,000	$\Phi = 8$ mm 4,500 $\Phi = 10$ and 12 mm 4,200 $\Phi = 14$ to 25 mm 4,000
Breaking stress (bar)	5,800	$\Phi \leq 20$ mm 4,850 $\Phi \geq 25$ mm 4,600	$\Phi \leq 20$ mm 4,850 $\Phi \geq 25$ mm 4,600	$\Phi \leq 16$ mm 5,000 $\Phi \geq 20$ mm 4,800	$\Phi = 8$ mm 5,500 $\Phi = 10$ and 12 mm 5,000 $\Phi = 14$ to 25 mm 4,800
Elongation at break (%)	21	14	14	12	12

High grip steel	Tentor steel	Tor steel	Caron steel	Nersid steel	Créloi steel
Yield strength taken into account in calculations (bar)	4,500	$\Phi \leq 20$ mm 4,200 $\Phi \geq 25$ mm 4,000	$\Phi \leq 20$ mm 4,200 $\Phi \geq 25$ mm 4,000	$\Phi \leq 16$ mm 4,200 $\Phi \geq 20$ mm 4,000	$\Phi = 8$ mm 4,500 $\Phi = 10$ and 12 mm 4,200 $\Phi = 14$–25 mm 4,000
Cracking coefficient	1.6	1.6	1.6	1.6	1.6
Sealing coefficient	1.414	1.414	1.414	1.414	1.414
Commercial lengths (m)	16	10–15	17	6–15	6–15
Diagram	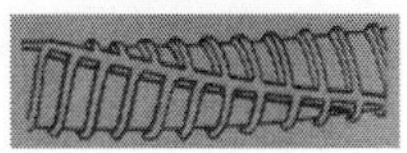		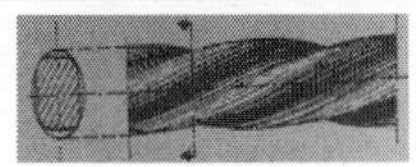		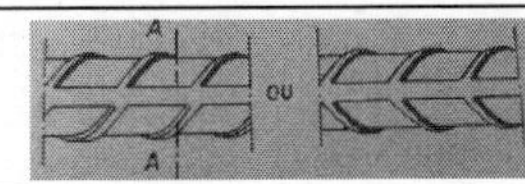
Manufacture	Mild steel hardened by traction and torsion	Torsionally hardened mild steel	Torsionally hardened mild steel	Naturally hard steel	Naturally hard steel

Table A1.3. *Summary table of welded mesh characteristics used according to various regulations*

Appendix 2

Characteristics of Concrete Used in the Various Regulations

1	Calculation method	Cement content or concrete class	Compressive stress of concrete during crushing (MPa)	Concrete compressive stress taken into account (MPa)
1906/1934	Allowable stresses	300 kg/m^3 350 kg/m^3 400 kg/m^3	16 18 20	4.4 5 5.6
1934/1945	Allowable stresses	300 kg/m^3 350 kg/m^3 400 kg/m^3	21.5 23.5 25	6 6.5 7
1945/1960	Allowable stresses	160–250 class cement, content: 350 kg 400 kg 250–315, class cement, content: 350 kg 400 kg		5 5.6 6.3 7.2

1960/1968	Allowable stresses	*Uncontrolled concrete* 250 kg/m^3 300 kg/m^3 350 kg/m^3 400 kg/m^3 *Controlled concrete* 250 kg/m^3 300 kg/m^3 350 kg/m^3 400 kg/m^3	 15 19 22.5 25 18 23 27 30	4.5 (CS) 9 (Fl.) 5.7 (CS) 11.4 (Fl.) 6.75 (CS) 13.5 (Fl.) 7.5 (CS) 15 (Fl.) 5.4 (CS) 10.8 (Fl.) 6.9 (CS) 13.8 (Fl.) 8.1 (CS) 16.2 (Fl.) 9 (CS) 18 (Fl.)
After 1968	Allowable stresses/limit states	Class B_1 Class B_2 Class B_3	25 30 35	According to the calculation rules

Table A2.1. *Summary table of material characteristics used according to various regulations*

NOTE.– CS: simple compression; Fl.: simple bending.

Appendix 3

Example of Calculations According to the Different Regulations

A3.1. Calculation of compressed parts likely to buckle

A rectangular column with a cross-section of 25 × 45 cm and a height of 8 m is considered. It is embedded at both ends and subjected to a compressive force of 75 t. The ratio of the live load to the design load (G/(G + Q)) is supposed to be 0.5. The compressive strength of the concrete is 65 kg/cm² (6.5 MPa).

A3.1.1. *Calculation according to the 1906 circular*

Article 12 of the Instruction validated the fact that parts whose slenderness was less than 20 and whose concrete compressive stress remained below the limit value were not to be checked for buckling. Otherwise, it was customary to use the Rankine formula:

$$N/\Omega\ (1 + kl^2/10{,}000\ r^2) \leq R_b$$

In our case, the slenderness is 800/25 = 32 ≥ 20, buckling must therefore be taken into account.

The value of k is 0.25 and l is the length of the column. The compression stress is then equal to:

$$75{,}000/(25 \times 45)\ (1 + 0/25 \times 800^2/(10.000 \times 25^2/12)) = 87.1\ \text{daN/cm}^2 \geq R_b = 65\ \text{kg/cm}^2$$

The post was underestimated according to the 1906 regulations (the reinforcements not being taken into account in the calculation).

A3.1.2. *Calculation according to the 1934 circular*

The slenderness of the column is 800/25 = 32 ≥ 24. The risk of buckling must therefore be taken into account. Applying the Rankine formula allows us to write:

$$N\,(1 + Kl^2/(10{,}000\ i^2)) = \sigma_b\,(B + nA)$$

with i = I/B, I being the inertia of the concrete section and B the concrete surface:

$$I = bh^3/12 = 45 \times 25^3/12, \text{ with } i^2 = 45 \times 25^3/(12 \times 25 \times 45) = 25^2/12 = 52.08$$

The 1934 rules gave a value of K = 1/4 for a double-embedded column. The formula can then be written as:

$$75{,}000\,(1 + 0.25 \times 800^2/(10{,}000 \times 52.08)) = 65\,(25 \times 45 + 10A)$$

We deduce that A = 38.30 cm² (4Φ25 + 4Φ25).

A3.1.3. *Calculation according to the 1945 circular*

Article 3 of the regulations then defined a "buckling unit f" such that $f = 4r\,\sqrt{E_b/n_j}$ as well as a buckling modulus $\lambda = l_f/f$. In this formula:

– r denotes the radius of gyration;

– E_b the elastic modulus of concrete: $E_b = 600\,(n_j)^{0,5} = 6{,}000\,(225)^{0.5} = 90{,}000$ bar;

– l_f the buckling length.

Depending on the values of λ obtained, the loads were multiplied by a buckling coefficient. This coefficient took the values from Table A3.1.

Values of λ	Values of the buckling coefficient
≤0.5	Buckling not taken into account
$0.5 \le \lambda \le 1$	2 λ
$1 \le \lambda \le 1.5$	$1 + \lambda^2$
$\lambda \ge 1.5$	$1.44\,\lambda^2$

Table A3.1. *Buckling coefficient values*

In our case, $f = 4/\sqrt{12} \times h \times \sqrt{E_b/n_j} = 6.8$ (for buckling the value of $E_b = 1.40\ E_v = 126{,}000$ bar). Here, $\lambda = l_f/f = 8/(2 \times 6.8) = 0.6$. Buckling must therefore be taken into account by multiplying the applied load by $2\ \lambda = 1.12$.

The calculation was carried out for the columns by checking that under the normal force alone, the admissible stresses of the materials were not exceeded. Under the force of 75 t, the compressive stress of the concrete is: $1.12 \times 75/(25 \times 45) = 75$ bar ≥ 65 bar. Since the permissible stress was exceeded, there was then the possibility of increasing the compressive strength of the concrete by hooping as stipulated in section 3.513 of the rules. In this case, the permissible stress for hooped concrete was given by:

$$(1 + \mu\ \omega_f\ (1 - 2e_f/a)\ n'_e/n_{90})\ R_b$$

with:

– ω_f: volume of transverse reinforcement (pins, frames, helical hoops, etc.) per unit volume of concrete;

– e_f: axis-to-axis spacing of transverse reinforcements;

– a: smallest cross-section of the column section;

– μ: shape coefficient of transverse reinforcements (see Table A3.2).

Shape of transverse reinforcements	Value of μ
Continuous helical frets or two layers of frames or pins	5.60
Square ligatures	2.80
Rectangular ligatures	2.80 a/b (a small dimension and b large dimension of the post)

Table A3.2. *Values of coefficient μ*

The value of the hoop coefficient was limited to 2.5. The compressive strength of the hooped section was then calculated with a surface area equal to:

$$\text{Min}\ (S_b + m\ \omega;\ 1.25\ S'_b + m\ \omega)$$

In this formula, S_b is the gross section of the concrete and S'_b the section of the fretted core. The minimum percentage of frets was 0.4%. In our example, the fretting coefficient considering square ligatures and a minimum percentage of 0.4% is 3.33 but limited to 2.5. The admissible stress then becomes $2.5 \times 65 = 162.5$ bar. With fretting, the compressive stress is lower than the admissible stress.

A3.1.4. *Calculation according to the 1960 circular*

The buckling length for a double-embedded post is equal to $l_c = l_0/2 = 8/2 = 4$ m. Buckling must be taken into account when the slenderness is greater than 50 (the concept of slenderness is different from that of the 1934 regulations). The limit value is $\lambda = l_c/i = 50$, which for a rectangular section of small side b amounts to verifying $l_c/b = 14.4$. The value is then $l_c/b = 4/0.25 = 16 \geq 14.4$, buckling must therefore be taken into account.

The radius of gyration is: $i = \sqrt{I/S} = 25/\sqrt{12} = 7.22$.

The elastic modulus of concrete E_b p for long-term loads is given by: $E_b = E_v = 7{,}000\ \sqrt{\sigma_j}$.

With: $\sigma_b = \alpha\beta\gamma\delta\varepsilon\ \sigma_j$, where $\sigma_j = \sigma_b/0.30 = 217$ bar.

This gives: $E_b = E_v = 7{,}000\ \sqrt{\sigma_j} = 7{,}000\ \sqrt{217} = 103{,}000$ bar.

Euler's critical stress is then: $\sigma_E = \pi^2\ E_b\ i^2/l_c^2 = 330$ bar.

And: $\sigma = 75{,}000/(25 \times 45) = 67$ bar $\leq 0.4 \times 330 = 132$ bar.

The column is therefore to be checked as a structure subjected to a normal force N and a bending moment $M = N \times f$ with:

$$f = \frac{1+\xi}{1300}\frac{(l_c - 50i)}{v}\frac{\sigma_E}{\sigma_E - \sigma}$$

In this formula, $\xi = 0.5 = G/(G + Q)$; v: distance from the center of gravity of the section to the contour = 25/2 = 12.5 cm. We then find:

$$F = (1 + 0.5)/1{,}300 \times (400 - 50 \times 7.22)^2/12.5 \times 330/(330 - 67) = 0.17 \text{ cm}$$

The bending moment to be taken into account is: $M = N \times f = 75 \times 0.0017 = 0.13$ tm.

A3.1.5. *Calculation according to the 1968 circular*

– Buckling check: $\lambda = l_f/i = 0.7 \times 8 \times \sqrt{12}/0.25 = 78 > 50$; buckling is therefore to be considered.

– Concrete compressive stress check: for the type of concrete chosen (350 kg/m^3 in attenuated control), the admissible stress is 68.5 bar. In addition, it is necessary to check that $\sigma'_b \leq \overline{\sigma}'_a/15 = 63$ bar.

The allowable stress for mild steels in FeE22 is given by: $\overline{\sigma}'_a = 2/3 \times 2{,}200 \times 2{,}160/3{,}340 = 950$ bar.

The minimum section of longitudinal reinforcement is: $A'_m = 1.25/1{,}000\ \theta_1\theta_2\theta_3\ N' / \overline{\sigma}'_{b0}$, so: for an interior column $\theta_1 = 1$, $\theta_2 = 1 + l_c/(4a - 2c) = 1 + 0.70 \times 800/(4 \times 25 - 2 \times 3) = 6.96$, $\theta_3 = 1 + 2{,}160/\sigma'_{en} = 1 + 2{,}160/2{,}160 = 2$. As a result: $A'_m = 1.25/1{,}000 \times 1 \times 6.96 \times 2 \times 75{,}000/68.5 = 19$ cm². The implemented section will therefore be 19 cm², and 4ϕ25 (19.63 cm²) with 8 mm frames every 30 cm.

The structure is calculated in composite bending with a fictitious bending moment $M = N \times f$, with $f = 8i^2/v\ (1 + \xi)\ 10^{-3}\ (\lambda - 50)^{3/2}$, with $\xi = G/(G + Q) = 0.5$ and $v = 0.25/2 = 0.125$ m.

Therefore $f = 0.07$ m, and $M = 75 \times 0.07 = 5.25$ tm.

A3.1.6. *Calculation according to the BAEL regulations*

– The slenderness is calculated as: $\lambda = l_f/i = 0.7 \times 8 \times \sqrt{12}/0.25 = 77 > 50$ (but less than 100).

– The minimum section of the reinforcements A_s is given as: Max (4 cm²/mL; 0.2% S_b) = Max (5.6 cm²; 2.25 cm²) = 5.6 cm².

– $\alpha = 0.60 \times (50/\lambda)^2 = 0.25$.

– $N_u \leq \alpha\ [(B_r \times f_{c28})/(0.9\ \gamma_b) + A_s \times f_e/1.15] = 51$ t with high-grip FeE40 steel and B25 concrete. This value is less than 1.4×75 t = 105 t.

The column had to be resized taking into account buckling (Faessel method, for example).

References

Charon, P. (1966). *Le calcul et la vérification des ouvrages en béton armé*. Éditions Eyrolles, Paris.

Charon, P. (1977). *Exercices de béton armé*. Éditions Eyrolles, Paris.

Debès, G. (1935). *Maçonneries, béton, béton armé*. Éditions Eyrolles, Paris.

Gauthron, R. (1957). *Calcul des cuves des châteaux d'eau*. Éditions Eyrolles, Paris.

Rabut, C. (1910–1911). *Cours de construction béton armé*. École nationale des ponts et chaussées, Paris.

Reimbert, M. and Reimbert, A. (1959). *Silos*. Éditions Eyrolles, Paris.

Codes, guides and standards

Building Code Requirements for Structural Concrete of 2018 (ACI 318M-18).

Cahier des charges général du 29 octobre 1913 intéressant le béton armé.

EN 1992-1-1. Eurocode 2: Design of concrete structures – Part 1-1: General rules and rules for buildings.

Fascicule 74 du CCTG. Construction des réservoirs en béton.

Instructions ministérielles du 20 octobre 1906 relatives à l'emploi du béton armé.

Instructions relatives à l'emploi du béton armé dans les ouvrages dépendant du ministère des travaux publics : circulaire série No. 8 du 19 juillet 1934.

International recommendations CEB-FIP 1970.

Note technique de janvier 1913 relative au calcul des ouvrages en ciment armé (section technique du génie militaire).

Règlement sur les constructions en béton armé établi par la chambre syndicale des constructeurs de ciment armé de France (22 janvier 1930).

Règles d'utilisation du béton armé applicables aux travaux dépendant du Ministère de la reconstruction et de l'Urbanisme et aux travaux privés de décembre 1945 (règles BA45 modifiées en mars 1948).

Règles Parasismiques PS 1969 (révisées 1982).

Règles Parasismiques PS 1992.

Règles pour le calcul et l'exécution des constructions en béton armé d'Octobre 1960 (règles BA60 révisées en 1964).

Règles techniques de conception des ouvrages et constructions en béton armé suivant la méthode des états limites du 7 octobre 1983 (règles BAEL 83).

Règles techniques de conception des ouvrages et constructions en béton armé suivant la méthode des états limites du 1 décembre 1992 (règles BAEL 91).

Règles techniques de conception et de calcul des ouvrages et construction en béton armé suivant la méthode des états limites, fascicule spécial n° 79-48 bis du 23 mai 1980 (règles BAEL 80).

Règles techniques de conception et de calcul des ouvrages et des constructions en béton armé de décembre 1968 (règles CCBA68).

Russian standard SP63.13330.2018. Concrete and reinforced concrete structures.

Index

E, M

S, T

Other titles from

iSTE

in

Civil Engineering and Geomechanics

2023

BRUNETTI Gian Luca
Design and Construction of Bioclimatic Wooden Greenhouses 1: Preliminary Design
Design and Construction of Bioclimatic Wooden Greenhouses 2: Design of Construction: Structure and Systems
Design and Construction of Bioclimatic Wooden Greenhouses 3: Design of Construction: Envelopes
Design and Construction of Bioclimatic Wooden Greenhouses 4: Architectural Integration and Quantitative Analyses

DHOUIB Ammar
Applied Geotechnics for Construction Projects 1: Soil and Experimental Data
Applied Geotechnics for Construction Projects 2: Fundamental Principles of Soil Mechanics and the Role of Water
Applied Geotechnics for Construction Projects 3: Behavior and Design of Project Foundations and Eurocode Validation
Applied Geotechnics for Construction Projects 4: Retaining Structures, Soil Reinforcement and Improvement

2022

MAZARS Jacky, GRANGE Stéphane
Damage and Cracking of Concrete Structures: From Theory to Practice

2020

SALENÇON Jean
Elastoplastic Modeling

2019

KOTRONIS Panagiotis
Risk Evaluation and Climate Change Adaptation of Civil Engineering Infrastructures and Buildings: Project RI-ADAPTCLIM

LAMBERT David Edward, PASILIAO Crystal L., ERZAR Benjamin, REVIL-BAUDARD Benoit, CAZACU Oana
Dynamic Damage and Fragmentation

PERROT Arnaud
3D Printing of Concrete: State of the Art and Challenges of the Digital Construction Revolution

SALENÇON Jean
Viscoelastic Modeling for Structural Analysis

ZEMBRI-MARY Geneviève
Project Risks: Actions Around Uncertainty in Urban Planning and Infrastructure Development

2018

FROSSARD Etienne
Granular Geomaterials Dissipative Mechanics: Theory and Applications in Civil Engineering

KHALFALLAH Salah
Structural Analysis 1: Statically Determinate Structures
Structural Analysis 2: Statically Indeterminate Structures

VERBRUGGE Jean-Claude, SCHROEDER Christian
Geotechnical Correlations for Soils and Rocks

2017

LAUZIN Xavier
Civil Engineering Structures According to the Eurocodes

PUECH Alain, GARNIER Jacques
Design of Piles Under Cyclic Loading: SOLCYP Recommendations

SELLIER Alain, GRIMAL Étienne, MULTON Stéphane, BOURDAROT Éric
Swelling Concrete in Dams and Hydraulic Structures: DSC 2017

2016

BARRE Francis, BISCH Philippe, CHAUVEL Danièle, CORTADE Jacques, COSTE Jean-François, DUBOIS Jean-Philippe, ERLICHER Silvano, GALLITRE Etienne, MAZARS Jacky, ROSPAR Claude, SELLIER Alain, TORRENTI Jean-Michel, TOUTLEMONDE François
Control of Cracking in Reinforced Concrete Structures: Research Project CEOS.fr

PIJAUDIER-CABOT Gilles, LA BORDERIE Christian, REESS Thierry, CHEN Wen, MAUREL Olivier, REY-BETHBEDER Franck, DE FERRON Antoine
Electrohydraulic Fracturing of Rocks

TORRENTI Jean-Michel, LA TORRE Francesca
Materials and Infrastructures 1
(Research for Innovative Transports Set – Volume 5A)
Materials and Infrastructures 2
(Research for Innovative Transports Set – Volume 5B)

2015

AÏT-MOKHTAR Abdelkarim, MILLET Olivier
Structure Design and Degradation Mechanisms in Coastal Environments

MONNET Jacques
In Situ *Tests in Geotechnical Engineering*

2014

DAÏAN Jean-François
Equilibrium and Transfer in Porous Media 1: Equilibrium States
Equilibrium and Transfer in Porous Media 2: Transfer Laws
Equilibrium and Transfer in Porous Media 3: Applications, Isothermal Transport and Coupled Transfers

2013

AMZIANE Sofiane, ARNAUD Laurent
Bio-aggregate-based Building Materials: Applications to Hemp Concretes

BONELLI Stéphane
Erosion in Geomechanics Applied to Dams and Levees

CASANDJIAN Charles, CHALLAMEL Noël, LANOS Christophe, HELLESLAND Jostein
Reinforced Concrete Beams, Columns and Frames: Mechanics and Design

GUÉGUEN Philippe
Seismic Vulnerability of Structures

HELLESLAND Jostein, CHALLAMEL Noël, CASANDJIAN Charles, LANOS Christophe
Reinforced Concrete Beams, Columns and Frames: Section and Slender Member Analysis

LALOUI Lyesse, DI DONNA Alice
Energy Geostructures: Innovation in Underground Engineering

2012

BONELLI Stéphane
Erosion of Geomaterials

JACOB Bernard, MCDONNELL Anne-Marie, SCHMIDT Franziska, CUNAGIN Wiley
ICWIM6 – Proceedings of the International Conference on Weigh-In-Motion

OLLIVIER Jean-Pierre, TORRENTI Jean-Marc, CARCASSÈS Myriam
Physical Properties of Concrete and Concrete Constituents

PIJAUDIER-CABOT Gilles, PEREIRA Jean-Michel
Geomechanics in CO_2 Storage Facilities

2011

BAROTH Julien, BREYSSE Denys, SCHOEFS Franck
Construction Reliability: Safety, Variability and Sustainability

CREMONA Christian
Structural Performance: Probability-based Assessment

HICHER Pierre-Yves
Multiscales Geomechanics: From Soil to Engineering Projects

IONESCU Ioan R., BOUVIER Salima, CAZACU Oana, FRANCIOSI Patrick,
Plasticity of Crystalline Materials: From Dislocations to Continuum

LOUKILI Ahmed
Self-Compacting Concrete

MOUTON Yves
Organic Materials for Sustainable Construction

LAMBERT Stéphane, NICOT François
Rockfall Engineering

PENSÉ-LHÉRITIER Anne-Marie
Formulation

PIJAUDIER-CABOT Gilles, DUFOUR Frédéric
Damage Mechanics of Cementitious Materials and Structures

RADJAÏ Farhang, DUBOIS Frédéric
Discrete-element Modeling of Granular Materials

TOUTLEMONDE François, RESPLENDINO Jacques
Designing and Building with UHPFRC: State-of-the-Art and Development

2010

ALSHIBLI Khalid A., REED Allen H.
Advances in Computed Tomography for Geomechanics: GeoX 2010

BUZAUD Eric, IONESCU Ioan R., VOYIADJIS Georges Z.
Materials under Extreme Loadings: Application to Penetration and Impact

LALOUI Lyesse
Mechanics of Unsaturated Geomechanics

NOVA Roberto
Soil Mechanics

SCHREFLER Bernard, DELAGE Pierre
Environmental Geomechanics

TORRENTI Jean-Michel, PIJAUDIER-CABOT Gilles, REYNOUARD Jean-Marie
Mechanical Behavior of Concrete

2009

AURIAULT Jean-Louis, BOUTIN Claude, GEINDREAU Christian
Homogenization of Coupled Phenomena in Heterogenous Media

CAMBOU Bernard, JEAN Michel, RADJAÏ Farhang
Micromechanics of Granular Materials

MAZARS Jacky, MILLARD Alain
Dynamic Behavior of Concrete and Seismic Engineering

NICOT François, WAN Richard
Micromechanics of Failure in Granular Geomechanics

2008

BETBEDER-MATIBET Jacques
Seismic Engineering

CAZACU Oana
Multiscale Modeling of Heterogenous Materials: From microstructure to macro-scale properties

HICHER Pierre-Yves, SHAO Jian-Fu
Constitutive Modeling of Soils and Rocks

JACOB Bernard, NORDENGEN Paul, O'CONNOR Alan, BOUTELDJA Mohamed
HVParis 2008 - HVTT10

JACOB Bernard, O'BRIEN Eugene, O'CONNOR Alan, BOUTELDJA Mohamed
HVParis 2008 - ICWIM5

SHAO Jian-Fu, BURLION Nicolas
Thermo-Hydromechanical and Chemical Coupling in Geomaterials and Applications: Proceedings of the 3rd International Symposium GeoProc'2008

2007

KUNDU Tribikram
Advanced Ultrasonic Methods for Material and Structure Inspection

2006

BALAGEAS Daniel, FRITZEN Claus-Peter, GÜEMES Alfredo
Structural Health Monitoring

DESRUES Jacques, VIGGIANI Gioacchino, BÉSUELLE Pierre
Advances in X-ray Tomography for Geomaterials

FSTT
Microtunneling and Horizontal Drilling: Recommendations

MOUTON Yves
Organic Materials in Civil Engineering

2005

PIJAUDIER-CABOT Gilles, GÉRARD Bruno, ACKER Paul
Creep Shrinkage and Durability of Concrete and Concrete Structures: CONCREEP – 7

Printed and bound by CPI Group (UK) Ltd, Croydon, CR0 4YY

26/08/2026

14949805-0001